MW00958335

THE Beekeeper's
ESSENTIAL
WORKBOOK

ALL-IN-ONE JOURNAL
TO TRACK AND ORGANIZE
YOUR BEEKEEPING SEASON

PIGGYBACK PRESS

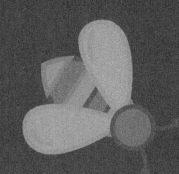

EXPENSES

EXPENSE LOG
ITEMS TO BUY

DATE	ITEM	VENDOR	QTY	AMOUNT
			TOTAL	

NOTES

EXPENSE LOG
ITEMS TO BUY

DATE	ITEM	VENDOR	QTY	AMOUNT
			TOTAL	

NOTES

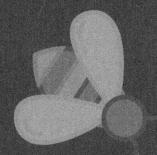

HIVE
INSPECTIONS

HIVE INSPECTION

Date: _____ Time: _____ Temp: _____

Weather Conditions: _____

Inspector: _____

Location: _____

NAME: _____

GENERAL APPEARANCE	HIVE#		HIVE#		HIVE#		HIVE#		HIVE#	
Hive Temperment	😖	😊	😖	😊	😖	😊	😖	😊	😖	😊
Bees actively entering and exiting the hive?	yes	no	yes	no	yes	no	yes	no	yes	no
Are bees bringing pollen?	yes	no	yes	no	yes	no	yes	no	yes	no
Are there signs of disturbance?	yes	no	yes	no	yes	no	yes	no	yes	no
How many boxes?										
OPENING THE HIVES										
Smoking material										
Capped honey	yes	no	yes	no	yes	no	yes	no	yes	no
Uncapped honey	yes	no	yes	no	yes	no	yes	no	yes	no
Capped brood	yes	no	yes	no	yes	no	yes	no	yes	no
Pollen	yes	no	yes	no	yes	no	yes	no	yes	no
Are there queen cells?	yes	no	yes	no	yes	no	yes	no	yes	no
Did you locate the queen?	yes	no	yes	no	yes	no	yes	no	yes	no
How many frames are covered in bees?										
Need space for nectar?	yes	no	yes	no	yes	no	yes	no	yes	no
PEST CONTROL Any signs of?										
Bald brood	yes	no	yes	no	yes	no	yes	no	yes	no
Foul brood	yes	no	yes	no	yes	no	yes	no	yes	no
Bad smell	yes	no	yes	no	yes	no	yes	no	yes	no
Moths	yes	no	yes	no	yes	no	yes	no	yes	no
Mice	yes	no	yes	no	yes	no	yes	no	yes	no
Ants	yes	no	yes	no	yes	no	yes	no	yes	no
Hive beetle	yes	no	yes	no	yes	no	yes	no	yes	no
Mold	yes	no	yes	no	yes	no	yes	no	yes	no
Varroa mites	yes	no	yes	no	yes	no	yes	no	yes	no
Other										
Overall Hive Health	1 2 3 4 5		1 2 3 4 5		1 2 3 4 5		1 2 3 4 5		1 2 3 4 5	

WHAT ACTIONS WERE TAKEN TODAY & WHAT IS TO BE SCHEDULED

HIVE INSPECTION

Date: _____ Time: _____ Temp: _____

Weather Conditions: _____

Inspector: _____ Location: _____

NAME: _____

GENERAL APPEARANCE	HIVE#		HIVE#		HIVE#		HIVE#		HIVE#	
Hive Temperment	😠	😊	😠	😊	😠	😊	😠	😊	😠	😊
Bees actively entering and exiting the hive?	yes	no	yes	no	yes	no	yes	no	yes	no
Are bees bringing pollen?	yes	no	yes	no	yes	no	yes	no	yes	no
Are there signs of disturbance?	yes	no	yes	no	yes	no	yes	no	yes	no
How many boxes?										

OPENING THE HIVES										
Smoking material										
Capped honey	yes	no	yes	no	yes	no	yes	no	yes	no
Uncapped honey	yes	no	yes	no	yes	no	yes	no	yes	no
Capped brood	yes	no	yes	no	yes	no	yes	no	yes	no
Pollen	yes	no	yes	no	yes	no	yes	no	yes	no
Are there queen cells?	yes	no	yes	no	yes	no	yes	no	yes	no
Did you locate the queen?	yes	no	yes	no	yes	no	yes	no	yes	no
How many frames are covered in bees?										
Need space for nectar?	yes	no	yes	no	yes	no	yes	no	yes	no

PEST CONTROL Any signs of?										
Bald brood	yes	no	yes	no	yes	no	yes	no	yes	no
Foul brood	yes	no	yes	no	yes	no	yes	no	yes	no
Bad smell	yes	no	yes	no	yes	no	yes	no	yes	no
Moths	yes	no	yes	no	yes	no	yes	no	yes	no
Mice	yes	no	yes	no	yes	no	yes	no	yes	no
Ants	yes	no	yes	no	yes	no	yes	no	yes	no
Hive beetle	yes	no	yes	no	yes	no	yes	no	yes	no
Mold	yes	no	yes	no	yes	no	yes	no	yes	no
Varroa mites	yes	no	yes	no	yes	no	yes	no	yes	no
Other										
Overall Hive Health	1 2 3 4 5		1 2 3 4 5		1 2 3 4 5		1 2 3 4 5		1 2 3 4 5	

WHAT ACTIONS WERE TAKEN TODAY & WHAT IS TO BE SCHEDULED

HIVE INSPECTION

Date: _____ Time: _____ Temp: _____

Weather Conditions: _____

Inspector: _____

Location: _____

NAME: _____

GENERAL APPEARANCE	HIVE#		HIVE#		HIVE#		HIVE#		HIVE#	
Hive Temperment	😠	😊	😠	😊	😠	😊	😠	😊	😠	😊
Bees actively entering and exiting the hive?	yes	no	yes	no	yes	no	yes	no	yes	no
Are bees bringing pollen?	yes	no	yes	no	yes	no	yes	no	yes	no
Are there signs of disturbance?	yes	no	yes	no	yes	no	yes	no	yes	no
How many boxes?										

OPENING THE HIVES

Smoking material										
Capped honey	yes	no	yes	no	yes	no	yes	no	yes	no
Uncapped honey	yes	no	yes	no	yes	no	yes	no	yes	no
Capped brood	yes	no	yes	no	yes	no	yes	no	yes	no
Pollen	yes	no	yes	no	yes	no	yes	no	yes	no
Are there queen cells?	yes	no	yes	no	yes	no	yes	no	yes	no
Did you locate the queen?	yes	no	yes	no	yes	no	yes	no	yes	no
How many frames are covered in bees?										
Need space for nectar?	yes	no	yes	no	yes	no	yes	no	yes	no

PEST CONTROL Any signs of?

Bald brood	yes	no	yes	no	yes	no	yes	no	yes	no
Foul brood	yes	no	yes	no	yes	no	yes	no	yes	no
Bad smell	yes	no	yes	no	yes	no	yes	no	yes	no
Moths	yes	no	yes	no	yes	no	yes	no	yes	no
Mice	yes	no	yes	no	yes	no	yes	no	yes	no
Ants	yes	no	yes	no	yes	no	yes	no	yes	no
Hive beetle	yes	no	yes	no	yes	no	yes	no	yes	no
Mold	yes	no	yes	no	yes	no	yes	no	yes	no
Varroa mites	yes	no	yes	no	yes	no	yes	no	yes	no
Other										
Overall Hive Health	1 2 3 4 5		1 2 3 4 5		1 2 3 4 5		1 2 3 4 5		1 2 3 4 5	

WHAT ACTIONS WERE TAKEN TODAY & WHAT IS TO BE SCHEDULED

HIVE INSPECTION

Date: _____ Time: _____ Temp: _____

Weather Conditions: _____

Inspector: _____

Location: _____

NAME: _____

GENERAL APPEARANCE	HIVE#		HIVE#		HIVE#		HIVE#		HIVE#	
Hive Temperment	😣	😊	😣	😊	😣	😊	😣	😊	😣	😊
Bees actively entering and exiting the hive?	yes	no	yes	no	yes	no	yes	no	yes	no
Are bees bringing pollen?	yes	no	yes	no	yes	no	yes	no	yes	no
Are there signs of disturbance?	yes	no	yes	no	yes	no	yes	no	yes	no
How many boxes?										

OPENING THE HIVES										
Smoking material										
Capped honey	yes	no	yes	no	yes	no	yes	no	yes	no
Uncapped honey	yes	no	yes	no	yes	no	yes	no	yes	no
Capped brood	yes	no	yes	no	yes	no	yes	no	yes	no
Pollen	yes	no	yes	no	yes	no	yes	no	yes	no
Are there queen cells?	yes	no	yes	no	yes	no	yes	no	yes	no
Did you locate the queen?	yes	no	yes	no	yes	no	yes	no	yes	no
How many frames are covered in bees?										
Need space for nectar?	yes	no	yes	no	yes	no	yes	no	yes	no

PEST CONTROL Any signs of?

Bald brood	yes	no	yes	no	yes	no	yes	no	yes	no
Foul brood	yes	no	yes	no	yes	no	yes	no	yes	no
Bad smell	yes	no	yes	no	yes	no	yes	no	yes	no
Moths	yes	no	yes	no	yes	no	yes	no	yes	no
Mice	yes	no	yes	no	yes	no	yes	no	yes	no
Ants	yes	no	yes	no	yes	no	yes	no	yes	no
Hive beetle	yes	no	yes	no	yes	no	yes	no	yes	no
Mold	yes	no	yes	no	yes	no	yes	no	yes	no
Varroa mites	yes	no	yes	no	yes	no	yes	no	yes	no
Other										
Overall Hive Health	1 2 3 4 5		1 2 3 4 5		1 2 3 4 5		1 2 3 4 5		1 2 3 4 5	

WHAT ACTIONS WERE TAKEN TODAY & WHAT IS TO BE SCHEDULED

_____

_____

_____

_____

_____

_____

_____

HIVE INSPECTION

Date: _____ Time: _____ Temp: _____

Weather Conditions: _____

Inspector: _____

Location: _____

NAME: _____

GENERAL APPEARANCE	HIVE#		HIVE#		HIVE#		HIVE#		HIVE#	
Hive Temperment	☹	☺	☹	☺	☹	☺	☹	☺	☹	☺
Bees actively entering and exiting the hive?	yes	no	yes	no	yes	no	yes	no	yes	no
Are bees bringing pollen?	yes	no	yes	no	yes	no	yes	no	yes	no
Are there signs of disturbance?	yes	no	yes	no	yes	no	yes	no	yes	no
How many boxes?										

OPENING THE HIVES										
Smoking material										
Capped honey	yes	no	yes	no	yes	no	yes	no	yes	no
Uncapped honey	yes	no	yes	no	yes	no	yes	no	yes	no
Capped brood	yes	no	yes	no	yes	no	yes	no	yes	no
Pollen	yes	no	yes	no	yes	no	yes	no	yes	no
Are there queen cells?	yes	no	yes	no	yes	no	yes	no	yes	no
Did you locate the queen?	yes	no	yes	no	yes	no	yes	no	yes	no
How many frames are covered in bees?										
Need space for nectar?	yes	no	yes	no	yes	no	yes	no	yes	no

PEST CONTROL Any signs of?										
Bald brood	yes	no	yes	no	yes	no	yes	no	yes	no
Foul brood	yes	no	yes	no	yes	no	yes	no	yes	no
Bad smell	yes	no	yes	no	yes	no	yes	no	yes	no
Moths	yes	no	yes	no	yes	no	yes	no	yes	no
Mice	yes	no	yes	no	yes	no	yes	no	yes	no
Ants	yes	no	yes	no	yes	no	yes	no	yes	no
Hive beetle	yes	no	yes	no	yes	no	yes	no	yes	no
Mold	yes	no	yes	no	yes	no	yes	no	yes	no
Varroa mites	yes	no	yes	no	yes	no	yes	no	yes	no
Other										
Overall Hive Health	1 2 3 4 5		1 2 3 4 5		1 2 3 4 5		1 2 3 4 5		1 2 3 4 5	

WHAT ACTIONS WERE TAKEN TODAY & WHAT IS TO BE SCHEDULED

HIVE INSPECTION

Date: _____ Time: _____ Temp: _____

Weather Conditions: _____

Inspector: _____

Location: _____

NAME: _____

GENERAL APPEARANCE	HIVE#		HIVE#		HIVE#		HIVE#		HIVE#	
Hive Temperment	☹	☺	☹	☺	☹	☺	☹	☺	☹	☺
Bees actively entering and exiting the hive?	yes	no	yes	no	yes	no	yes	no	yes	no
Are bees bringing pollen?	yes	no	yes	no	yes	no	yes	no	yes	no
Are there signs of disturbance?	yes	no	yes	no	yes	no	yes	no	yes	no
How many boxes?										

OPENING THE HIVES

Smoking material										
Capped honey	yes	no	yes	no	yes	no	yes	no	yes	no
Uncapped honey	yes	no	yes	no	yes	no	yes	no	yes	no
Capped brood	yes	no	yes	no	yes	no	yes	no	yes	no
Pollen	yes	no	yes	no	yes	no	yes	no	yes	no
Are there queen cells?	yes	no	yes	no	yes	no	yes	no	yes	no
Did you locate the queen?	yes	no	yes	no	yes	no	yes	no	yes	no
How many frames are covered in bees?										
Need space for nectar?	yes	no	yes	no	yes	no	yes	no	yes	no

PEST CONTROL Any signs of?

Bald brood	yes	no	yes	no	yes	no	yes	no	yes	no
Foul brood	yes	no	yes	no	yes	no	yes	no	yes	no
Bad smell	yes	no	yes	no	yes	no	yes	no	yes	no
Moths	yes	no	yes	no	yes	no	yes	no	yes	no
Mice	yes	no	yes	no	yes	no	yes	no	yes	no
Ants	yes	no	yes	no	yes	no	yes	no	yes	no
Hive beetle	yes	no	yes	no	yes	no	yes	no	yes	no
Mold	yes	no	yes	no	yes	no	yes	no	yes	no
Varroa mites	yes	no	yes	no	yes	no	yes	no	yes	no
Other										
Overall Hive Health	1 2 3 4 5		1 2 3 4 5		1 2 3 4 5		1 2 3 4 5		1 2 3 4	

WHAT ACTIONS WERE TAKEN TODAY & WHAT IS TO BE SCHEDULED

HIVE INSPECTION

Date: _____ Time: _____ Temp: _____

Weather Conditions: _____

Inspector: _____ Location: _____

NAME: _____

GENERAL APPEARANCE	HIVE#		HIVE#		HIVE#		HIVE#		HIVE#	
Hive Temperment	😠	😊	😠	😊	😠	😊	😠	😊	😠	😊
Bees actively entering and exiting the hive?	yes	no	yes	no	yes	no	yes	no	yes	no
Are bees bringing pollen?	yes	no	yes	no	yes	no	yes	no	yes	no
Are there signs of disturbance?	yes	no	yes	no	yes	no	yes	no	yes	no
How many boxes?										
OPENING THE HIVES										
Smoking material										
Capped honey	yes	no	yes	no	yes	no	yes	no	yes	no
Uncapped honey	yes	no	yes	no	yes	no	yes	no	yes	no
Capped brood	yes	no	yes	no	yes	no	yes	no	yes	no
Pollen	yes	no	yes	no	yes	no	yes	no	yes	no
Are there queen cells?	yes	no	yes	no	yes	no	yes	no	yes	no
Did you locate the queen?	yes	no	yes	no	yes	no	yes	no	yes	no
How many frames are covered in bees?										
Need space for nectar?	yes	no	yes	no	yes	no	yes	no	yes	no
PEST CONTROL Any signs of?										
Bald brood	yes	no	yes	no	yes	no	yes	no	yes	no
Foul brood	yes	no	yes	no	yes	no	yes	no	yes	no
Bad smell	yes	no	yes	no	yes	no	yes	no	yes	no
Moths	yes	no	yes	no	yes	no	yes	no	yes	no
Mice	yes	no	yes	no	yes	no	yes	no	yes	no
Ants	yes	no	yes	no	yes	no	yes	no	yes	no
Hive beetle	yes	no	yes	no	yes	no	yes	no	yes	no
Mold	yes	no	yes	no	yes	no	yes	no	yes	no
Varroa mites	yes	no	yes	no	yes	no	yes	no	yes	no
Other										
Overall Hive Health	1 2 3 4 5		1 2 3 4 5		1 2 3 4 5		1 2 3 4 5		1 2 3 4 5	

WHAT ACTIONS WERE TAKEN TODAY & WHAT IS TO BE SCHEDULED

HIVE INSPECTION

Date: _____ Time: _____ Temp: _____

Weather Conditions: _____

Inspector: _____ Location: _____

NAME: _____

GENERAL APPEARANCE	HIVE#		HIVE#		HIVE#		HIVE#		HIVE#	
Hive Temperment	☹	☺	☹	☺	☹	☺	☹	☺	☹	☺
Bees actively entering and exiting the hive?	yes	no	yes	no	yes	no	yes	no	yes	no
Are bees bringing pollen?	yes	no	yes	no	yes	no	yes	no	yes	no
Are there signs of disturbance?	yes	no	yes	no	yes	no	yes	no	yes	no
How many boxes?										

OPENING THE HIVES										
Smoking material										
Capped honey	yes	no	yes	no	yes	no	yes	no	yes	no
Uncapped honey	yes	no	yes	no	yes	no	yes	no	yes	no
Capped brood	yes	no	yes	no	yes	no	yes	no	yes	no
Pollen	yes	no	yes	no	yes	no	yes	no	yes	no
Are there queen cells?	yes	no	yes	no	yes	no	yes	no	yes	no
Did you locate the queen?	yes	no	yes	no	yes	no	yes	no	yes	no
How many frames are covered in bees?										
Need space for nectar?	yes	no	yes	no	yes	no	yes	no	yes	no

PEST CONTROL Any signs of?										
Bald brood	yes	no	yes	no	yes	no	yes	no	yes	no
Foul brood	yes	no	yes	no	yes	no	yes	no	yes	no
Bad smell	yes	no	yes	no	yes	no	yes	no	yes	no
Moths	yes	no	yes	no	yes	no	yes	no	yes	no
Mice	yes	no	yes	no	yes	no	yes	no	yes	no
Ants	yes	no	yes	no	yes	no	yes	no	yes	no
Hive beetle	yes	no	yes	no	yes	no	yes	no	yes	no
Mold	yes	no	yes	no	yes	no	yes	no	yes	no
Varroa mites	yes	no	yes	no	yes	no	yes	no	yes	no
Other										
Overall Hive Health	1 2 3 4 5		1 2 3 4 5		1 2 3 4 5		1 2 3 4 5		1 2 3 4 5	

WHAT ACTIONS WERE TAKEN TODAY & WHAT IS TO BE SCHEDULED

HIVE INSPECTION

Date: _____ Time: _____ Temp: _____

Weather Conditions: _____

Inspector: _____ Location: _____

NAME: _____

GENERAL APPEARANCE	HIVE#		HIVE#		HIVE#		HIVE#		HIVE#	
Hive Temperment	😖	😊	😖	😊	😖	😊	😖	😊	😖	😊
Bees actively entering and exiting the hive?	yes	no	yes	no	yes	no	yes	no	yes	no
Are bees bringing pollen?	yes	no	yes	no	yes	no	yes	no	yes	no
Are there signs of disturbance?	yes	no	yes	no	yes	no	yes	no	yes	no
How many boxes?										

OPENING THE HIVES										
Smoking material										
Capped honey	yes	no	yes	no	yes	no	yes	no	yes	no
Uncapped honey	yes	no	yes	no	yes	no	yes	no	yes	no
Capped brood	yes	no	yes	no	yes	no	yes	no	yes	no
Pollen	yes	no	yes	no	yes	no	yes	no	yes	no
Are there queen cells?	yes	no	yes	no	yes	no	yes	no	yes	no
Did you locate the queen?	yes	no	yes	no	yes	no	yes	no	yes	no
How many frames are covered in bees?										
Need space for nectar?	yes	no	yes	no	yes	no	yes	no	yes	no

PEST CONTROL Any signs of?										
Bald brood	yes	no	yes	no	yes	no	yes	no	yes	no
Foul brood	yes	no	yes	no	yes	no	yes	no	yes	no
Bad smell	yes	no	yes	no	yes	no	yes	no	yes	no
Moths	yes	no	yes	no	yes	no	yes	no	yes	no
Mice	yes	no	yes	no	yes	no	yes	no	yes	no
Ants	yes	no	yes	no	yes	no	yes	no	yes	no
Hive beetle	yes	no	yes	no	yes	no	yes	no	yes	no
Mold	yes	no	yes	no	yes	no	yes	no	yes	no
Varroa mites	yes	no	yes	no	yes	no	yes	no	yes	no
Other										
Overall Hive Health	1 2 3 4 5		1 2 3 4 5		1 2 3 4 5		1 2 3 4 5		1 2 3 4 5	

WHAT ACTIONS WERE TAKEN TODAY & WHAT IS TO BE SCHEDULED

HIVE INSPECTION

Date: .... Time: .... Temp: ....

Weather Conditions: ....

Inspector: .... Location: ....

NAME: ....

GENERAL APPEARANCE	HIVE#		HIVE#		HIVE#		HIVE#		HIVE#	
Hive Temperment	☹	☺	☹	☺	☹	☺	☹	☺	☹	☺
Bees actively entering and exiting the hive?	yes	no	yes	no	yes	no	yes	no	yes	no
Are bees bringing pollen?	yes	no	yes	no	yes	no	yes	no	yes	no
Are there signs of disturbance?	yes	no	yes	no	yes	no	yes	no	yes	no
How many boxes?										

OPENING THE HIVES										
Smoking material										
Capped honey	yes	no	yes	no	yes	no	yes	no	yes	no
Uncapped honey	yes	no	yes	no	yes	no	yes	no	yes	no
Capped brood	yes	no	yes	no	yes	no	yes	no	yes	no
Pollen	yes	no	yes	no	yes	no	yes	no	yes	no
Are there queen cells?	yes	no	yes	no	yes	no	yes	no	yes	no
Did you locate the queen?	yes	no	yes	no	yes	no	yes	no	yes	no
How many frames are covered in bees?										
Need space for nectar?	yes	no	yes	no	yes	no	yes	no	yes	no

PEST CONTROL Any signs of?										
Bald brood	yes	no	yes	no	yes	no	yes	no	yes	no
Foul brood	yes	no	yes	no	yes	no	yes	no	yes	no
Bad smell	yes	no	yes	no	yes	no	yes	no	yes	no
Moths	yes	no	yes	no	yes	no	yes	no	yes	no
Mice	yes	no	yes	no	yes	no	yes	no	yes	no
Ants	yes	no	yes	no	yes	no	yes	no	yes	no
Hive beetle	yes	no	yes	no	yes	no	yes	no	yes	no
Mold	yes	no	yes	no	yes	no	yes	no	yes	no
Varroa mites	yes	no	yes	no	yes	no	yes	no	yes	no
Other										
Overall Hive Health	1 2 3 4 5		1 2 3 4 5		1 2 3 4 5		1 2 3 4 5		1 2 3 4 5	

WHAT ACTIONS WERE TAKEN TODAY & WHAT IS TO BE SCHEDULED

HIVE INSPECTION

Date: _____ Time: _____ Temp: _____

Weather Conditions: _____

Inspector: _____ Location: _____

NAME: _____

GENERAL APPEARANCE	HIVE#		HIVE#		HIVE#		HIVE#		HIVE#	
Hive Temperment	😠	😊	😠	😊	😠	😊	😠	😊	😠	😊
Bees actively entering and exiting the hive?	yes	no	yes	no	yes	no	yes	no	yes	no
Are bees bringing pollen?	yes	no	yes	no	yes	no	yes	no	yes	no
Are there signs of disturbance?	yes	no	yes	no	yes	no	yes	no	yes	no
How many boxes?										

OPENING THE HIVES										
Smoking material										
Capped honey	yes	no	yes	no	yes	no	yes	no	yes	no
Uncapped honey	yes	no	yes	no	yes	no	yes	no	yes	no
Capped brood	yes	no	yes	no	yes	no	yes	no	yes	no
Pollen	yes	no	yes	no	yes	no	yes	no	yes	no
Are there queen cells?	yes	no	yes	no	yes	no	yes	no	yes	no
Did you locate the queen?	yes	no	yes	no	yes	no	yes	no	yes	no
How many frames are covered in bees?										
Need space for nectar?	yes	no	yes	no	yes	no	yes	no	yes	no

PEST CONTROL Any signs of?										
Bald brood	yes	no	yes	no	yes	no	yes	no	yes	no
Foul brood	yes	no	yes	no	yes	no	yes	no	yes	no
Bad smell	yes	no	yes	no	yes	no	yes	no	yes	no
Moths	yes	no	yes	no	yes	no	yes	no	yes	no
Mice	yes	no	yes	no	yes	no	yes	no	yes	no
Ants	yes	no	yes	no	yes	no	yes	no	yes	no
Hive beetle	yes	no	yes	no	yes	no	yes	no	yes	no
Mold	yes	no	yes	no	yes	no	yes	no	yes	no
Varroa mites	yes	no	yes	no	yes	no	yes	no	yes	no
Other										
Overall Hive Health	1 2 3 4 5		1 2 3 4 5		1 2 3 4 5		1 2 3 4 5		1 2 3 4 5	

WHAT ACTIONS WERE TAKEN TODAY & WHAT IS TO BE SCHEDULED

HIVE INSPECTION

Date: _____ Time: _____ Temp: _____

Weather Conditions: _____

Inspector: _____ Location: _____

NAME:

GENERAL APPEARANCE	HIVE#		HIVE#		HIVE#		HIVE#		HIVE#	
Hive Temperment	😠	😊	😠	😊	😠	😊	😠	😊	😠	😊
Bees actively entering and exiting the hive?	yes	no	yes	no	yes	no	yes	no	yes	no
Are bees bringing pollen?	yes	no	yes	no	yes	no	yes	no	yes	no
Are there signs of disturbance?	yes	no	yes	no	yes	no	yes	no	yes	no
How many boxes?										

OPENING THE HIVES										
Smoking material										
Capped honey	yes	no	yes	no	yes	no	yes	no	yes	no
Uncapped honey	yes	no	yes	no	yes	no	yes	no	yes	no
Capped brood	yes	no	yes	no	yes	no	yes	no	yes	no
Pollen	yes	no	yes	no	yes	no	yes	no	yes	no
Are there queen cells?	yes	no	yes	no	yes	no	yes	no	yes	no
Did you locate the queen?	yes	no	yes	no	yes	no	yes	no	yes	no
How many frames are covered in bees?										
Need space for nectar?	yes	no	yes	no	yes	no	yes	no	yes	no

PEST CONTROL Any signs of?										
Bald brood	yes	no	yes	no	yes	no	yes	no	yes	no
Foul brood	yes	no	yes	no	yes	no	yes	no	yes	no
Bad smell	yes	no	yes	no	yes	no	yes	no	yes	no
Moths	yes	no	yes	no	yes	no	yes	no	yes	no
Mice	yes	no	yes	no	yes	no	yes	no	yes	no
Ants	yes	no	yes	no	yes	no	yes	no	yes	no
Hive beetle	yes	no	yes	no	yes	no	yes	no	yes	no
Mold	yes	no	yes	no	yes	no	yes	no	yes	no
Varroa mites	yes	no	yes	no	yes	no	yes	no	yes	no
Other										
Overall Hive Health	1 2 3 4 5		1 2 3 4 5		1 2 3 4 5		1 2 3 4 5		1 2 3 4 5	

WHAT ACTIONS WERE TAKEN TODAY & WHAT IS TO BE SCHEDULED

HIVE INSPECTION

Date: _____ Time: _____ Temp: _____

Weather Conditions: _____

Inspector: _____ Location: _____

NAME: _____

GENERAL APPEARANCE	HIVE#		HIVE#		HIVE#		HIVE#		HIVE#	
Hive Temperment	☹	☺	☹	☺	☹	☺	☹	☺	☹	☺
Bees actively entering and exiting the hive?	yes	no	yes	no	yes	no	yes	no	yes	no
Are bees bringing pollen?	yes	no	yes	no	yes	no	yes	no	yes	no
Are there signs of disturbance?	yes	no	yes	no	yes	no	yes	no	yes	no
How many boxes?										

OPENING THE HIVES

Smoking material										
Capped honey	yes	no	yes	no	yes	no	yes	no	yes	no
Uncapped honey	yes	no	yes	no	yes	no	yes	no	yes	no
Capped brood	yes	no	yes	no	yes	no	yes	no	yes	no
Pollen	yes	no	yes	no	yes	no	yes	no	yes	no
Are there queen cells?	yes	no	yes	no	yes	no	yes	no	yes	no
Did you locate the queen?	yes	no	yes	no	yes	no	yes	no	yes	no
How many frames are covered in bees?										
Need space for nectar?	yes	no	yes	no	yes	no	yes	no	yes	no

PEST CONTROL Any signs of?

Bald brood	yes	no	yes	no	yes	no	yes	no	yes	no
Foul brood	yes	no	yes	no	yes	no	yes	no	yes	no
Bad smell	yes	no	yes	no	yes	no	yes	no	yes	no
Moths	yes	no	yes	no	yes	no	yes	no	yes	no
Mice	yes	no	yes	no	yes	no	yes	no	yes	no
Ants	yes	no	yes	no	yes	no	yes	no	yes	no
Hive beetle	yes	no	yes	no	yes	no	yes	no	yes	no
Mold	yes	no	yes	no	yes	no	yes	no	yes	no
Varroa mites	yes	no	yes	no	yes	no	yes	no	yes	no
Other										
Overall Hive Health	1 2 3 4 5		1 2 3 4 5		1 2 3 4 5		1 2 3 4 5		1 2 3 4 5	

WHAT ACTIONS WERE TAKEN TODAY & WHAT IS TO BE SCHEDULED

_____

_____

_____

_____

_____

_____

_____

_____

_____

HIVE INSPECTION

Date: _____ Time: _____ Temp: _____

Weather Conditions: _____

Inspector: _____ Location: _____

NAME: _____

GENERAL APPEARANCE

	HIVE# ___		HIVE# ___		HIVE# ___		HIVE# ___		HIVE# ___	
Hive Temperment	☹	☺	☹	☺	☹	☺	☹	☺	☹	☺
Bees actively entering and exiting the hive?	yes	no	yes	no	yes	no	yes	no	yes	no
Are bees bringing pollen?	yes	no	yes	no	yes	no	yes	no	yes	no
Are there signs of disturbance?	yes	no	yes	no	yes	no	yes	no	yes	no
How many boxes?										

OPENING THE HIVES

Smoking material										
Capped honey	yes	no	yes	no	yes	no	yes	no	yes	no
Uncapped honey	yes	no	yes	no	yes	no	yes	no	yes	no
Capped brood	yes	no	yes	no	yes	no	yes	no	yes	no
Pollen	yes	no	yes	no	yes	no	yes	no	yes	no
Are there queen cells?	yes	no	yes	no	yes	no	yes	no	yes	no
Did you locate the queen?	yes	no	yes	no	yes	no	yes	no	yes	no
How many frames are covered in bees?										
Need space for nectar?	yes	no	yes	no	yes	no	yes	no	yes	no

PEST CONTROL Any signs of?

Bald brood	yes	no	yes	no	yes	no	yes	no	yes	no
Foul brood	yes	no	yes	no	yes	no	yes	no	yes	no
Bad smell	yes	no	yes	no	yes	no	yes	no	yes	no
Moths	yes	no	yes	no	yes	no	yes	no	yes	no
Mice	yes	no	yes	no	yes	no	yes	no	yes	no
Ants	yes	no	yes	no	yes	no	yes	no	yes	no
Hive beetle	yes	no	yes	no	yes	no	yes	no	yes	no
Mold	yes	no	yes	no	yes	no	yes	no	yes	no
Varroa mites	yes	no	yes	no	yes	no	yes	no	yes	no
Other										
Overall Hive Health	1 2 3 4 5		1 2 3 4 5		1 2 3 4 5		1 2 3 4 5		1 2 3 4 5	

WHAT ACTIONS WERE TAKEN TODAY & WHAT IS TO BE SCHEDULED

HIVE INSPECTION

Date: Time: Temp:

Weather Conditions:

Inspector: Location:

NAME:

GENERAL APPEARANCE	HIVE#		HIVE#		HIVE#		HIVE#		HIVE#	
Hive Temperment	☹	☺	☹	☺	☹	☺	☹	☺	☹	☺
Bees actively entering and exiting the hive?	yes	no	yes	no	yes	no	yes	no	yes	no
Are bees bringing pollen?	yes	no	yes	no	yes	no	yes	no	yes	no
Are there signs of disturbance?	yes	no	yes	no	yes	no	yes	no	yes	no
How many boxes?										
OPENING THE HIVES										
Smoking material										
Capped honey	yes	no	yes	no	yes	no	yes	no	yes	no
Uncapped honey	yes	no	yes	no	yes	no	yes	no	yes	no
Capped brood	yes	no	yes	no	yes	no	yes	no	yes	no
Pollen	yes	no	yes	no	yes	no	yes	no	yes	no
Are there queen cells?	yes	no	yes	no	yes	no	yes	no	yes	no
Did you locate the queen?	yes	no	yes	no	yes	no	yes	no	yes	no
How many frames are covered in bees?										
Need space for nectar?	yes	no	yes	no	yes	no	yes	no	yes	no
PEST CONTROL Any signs of?										
Bald brood	yes	no	yes	no	yes	no	yes	no	yes	no
Foul brood	yes	no	yes	no	yes	no	yes	no	yes	no
Bad smell	yes	no	yes	no	yes	no	yes	no	yes	no
Moths	yes	no	yes	no	yes	no	yes	no	yes	no
Mice	yes	no	yes	no	yes	no	yes	no	yes	no
Ants	yes	no	yes	no	yes	no	yes	no	yes	no
Hive beetle	yes	no	yes	no	yes	no	yes	no	yes	no
Mold	yes	no	yes	no	yes	no	yes	no	yes	no
Varroa mites	yes	no	yes	no	yes	no	yes	no	yes	no
Other										
Overall Hive Health	1 2 3 4 5		1 2 3 4 5		1 2 3 4 5		1 2 3 4 5		1 2 3 4 5	

WHAT ACTIONS WERE TAKEN TODAY & WHAT IS TO BE SCHEDULED

HIVE INSPECTION

Date: _____ Time: _____ Temp: _____

Weather Conditions: _____

Inspector: _____

Location: _____

NAME:

GENERAL APPEARANCE	HIVE#		HIVE#		HIVE#		HIVE#		HIVE#	
Hive Temperment	☹	☺	☹	☺	☹	☺	☹	☺	☹	☺
Bees actively entering and exiting the hive?	yes	no	yes	no	yes	no	yes	no	yes	no
Are bees bringing pollen?	yes	no	yes	no	yes	no	yes	no	yes	no
Are there signs of disturbance?	yes	no	yes	no	yes	no	yes	no	yes	no
How many boxes?										

OPENING THE HIVES										
Smoking material										
Capped honey	yes	no	yes	no	yes	no	yes	no	yes	no
Uncapped honey	yes	no	yes	no	yes	no	yes	no	yes	no
Capped brood	yes	no	yes	no	yes	no	yes	no	yes	no
Pollen	yes	no	yes	no	yes	no	yes	no	yes	no
Are there queen cells?	yes	no	yes	no	yes	no	yes	no	yes	no
Did you locate the queen?	yes	no	yes	no	yes	no	yes	no	yes	no
How many frames are covered in bees?										
Need space for nectar?	yes	no	yes	no	yes	no	yes	no	yes	no

PEST CONTROL Any signs of?										
Bald brood	yes	no	yes	no	yes	no	yes	no	yes	no
Foul brood	yes	no	yes	no	yes	no	yes	no	yes	no
Bad smell	yes	no	yes	no	yes	no	yes	no	yes	no
Moths	yes	no	yes	no	yes	no	yes	no	yes	no
Mice	yes	no	yes	no	yes	no	yes	no	yes	no
Ants	yes	no	yes	no	yes	no	yes	no	yes	no
Hive beetle	yes	no	yes	no	yes	no	yes	no	yes	no
Mold	yes	no	yes	no	yes	no	yes	no	yes	no
Varroa mites	yes	no	yes	no	yes	no	yes	no	yes	no
Other										
Overall Hive Health	1 2 3 4 5		1 2 3 4 5		1 2 3 4 5		1 2 3 4 5		1 2 3 4 5	

WHAT ACTIONS WERE TAKEN TODAY & WHAT IS TO BE SCHEDULED

HIVE INSPECTION

Date: _____ Time: _____ Temp: _____

Weather Conditions: _____

Inspector: _____

Location: _____

NAME: _____

GENERAL APPEARANCE	HIVE#		HIVE#		HIVE#		HIVE#		HIVE#	
Hive Temperment	☹	☺	☹	☺	☹	☺	☹	☺	☹	☺
Bees actively entering and exiting the hive?	yes	no	yes	no	yes	no	yes	no	yes	no
Are bees bringing pollen?	yes	no	yes	no	yes	no	yes	no	yes	no
Are there signs of disturbance?	yes	no	yes	no	yes	no	yes	no	yes	no
How many boxes?										
OPENING THE HIVES										
Smoking material										
Capped honey	yes	no	yes	no	yes	no	yes	no	yes	no
Uncapped honey	yes	no	yes	no	yes	no	yes	no	yes	no
Capped brood	yes	no	yes	no	yes	no	yes	no	yes	no
Pollen	yes	no	yes	no	yes	no	yes	no	yes	no
Are there queen cells?	yes	no	yes	no	yes	no	yes	no	yes	no
Did you locate the queen?	yes	no	yes	no	yes	no	yes	no	yes	no
How many frames are covered in bees?										
Need space for nectar?	yes	no	yes	no	yes	no	yes	no	yes	no
PEST CONTROL Any signs of?										
Bald brood	yes	no	yes	no	yes	no	yes	no	yes	no
Foul brood	yes	no	yes	no	yes	no	yes	no	yes	no
Bad smell	yes	no	yes	no	yes	no	yes	no	yes	no
Moths	yes	no	yes	no	yes	no	yes	no	yes	no
Mice	yes	no	yes	no	yes	no	yes	no	yes	no
Ants	yes	no	yes	no	yes	no	yes	no	yes	no
Hive beetle	yes	no	yes	no	yes	no	yes	no	yes	no
Mold	yes	no	yes	no	yes	no	yes	no	yes	no
Varroa mites	yes	no	yes	no	yes	no	yes	no	yes	no
Other										
Overall Hive Health	1 2 3 4 5		1 2 3 4 5		1 2 3 4 5		1 2 3 4 5		1 2 3 4 5	

WHAT ACTIONS WERE TAKEN TODAY & WHAT IS TO BE SCHEDULED

HIVE INSPECTION

Date: _____ Time: _____ Temp: _____

Weather Conditions: _____

Inspector: _____ Location: _____

NAME: _____

GENERAL APPEARANCE	HIVE#		HIVE#		HIVE#		HIVE#		HIVE#	
Hive Temperment	😠	😊	😠	😊	😠	😊	😠	😊	😠	😊
Bees actively entering and exiting the hive?	yes	no	yes	no	yes	no	yes	no	yes	no
Are bees bringing pollen?	yes	no	yes	no	yes	no	yes	no	yes	no
Are there signs of disturbance?	yes	no	yes	no	yes	no	yes	no	yes	no
How many boxes?										

OPENING THE HIVES										
Smoking material										
Capped honey	yes	no	yes	no	yes	no	yes	no	yes	no
Uncapped honey	yes	no	yes	no	yes	no	yes	no	yes	no
Capped brood	yes	no	yes	no	yes	no	yes	no	yes	no
Pollen	yes	no	yes	no	yes	no	yes	no	yes	no
Are there queen cells?	yes	no	yes	no	yes	no	yes	no	yes	no
Did you locate the queen?	yes	no	yes	no	yes	no	yes	no	yes	no
How many frames are covered in bees?										
Need space for nectar?	yes	no	yes	no	yes	no	yes	no	yes	no

PEST CONTROL Any signs of?										
Bald brood	yes	no	yes	no	yes	no	yes	no	yes	no
Foul brood	yes	no	yes	no	yes	no	yes	no	yes	no
Bad smell	yes	no	yes	no	yes	no	yes	no	yes	no
Moths	yes	no	yes	no	yes	no	yes	no	yes	no
Mice	yes	no	yes	no	yes	no	yes	no	yes	no
Ants	yes	no	yes	no	yes	no	yes	no	yes	no
Hive beetle	yes	no	yes	no	yes	no	yes	no	yes	no
Mold	yes	no	yes	no	yes	no	yes	no	yes	no
Varroa mites	yes	no	yes	no	yes	no	yes	no	yes	no
Other										
Overall Hive Health	1 2 3 4 5		1 2 3 4 5		1 2 3 4 5		1 2 3 4 5		1 2 3 4 5	

WHAT ACTIONS WERE TAKEN TODAY & WHAT IS TO BE SCHEDULED

HIVE INSPECTION

Date: _____ Time: _____ Temp: _____

Weather Conditions: _____

Inspector: _____ Location: _____

NAME: _____

GENERAL APPEARANCE	HIVE#		HIVE#		HIVE#		HIVE#		HIVE#	
Hive Temperment	☹	☺	☹	☺	☹	☺	☹	☺	☹	☺
Bees actively entering and exiting the hive?	yes	no	yes	no	yes	no	yes	no	yes	no
Are bees bringing pollen?	yes	no	yes	no	yes	no	yes	no	yes	no
Are there signs of disturbance?	yes	no	yes	no	yes	no	yes	no	yes	no
How many boxes?										
OPENING THE HIVES										
Smoking material										
Capped honey	yes	no	yes	no	yes	no	yes	no	yes	no
Uncapped honey	yes	no	yes	no	yes	no	yes	no	yes	no
Capped brood	yes	no	yes	no	yes	no	yes	no	yes	no
Pollen	yes	no	yes	no	yes	no	yes	no	yes	no
Are there queen cells?	yes	no	yes	no	yes	no	yes	no	yes	no
Did you locate the queen?	yes	no	yes	no	yes	no	yes	no	yes	no
How many frames are covered in bees?										
Need space for nectar?	yes	no	yes	no	yes	no	yes	no	yes	no
PEST CONTROL Any signs of?										
Bald brood	yes	no	yes	no	yes	no	yes	no	yes	no
Foul brood	yes	no	yes	no	yes	no	yes	no	yes	no
Bad smell	yes	no	yes	no	yes	no	yes	no	yes	no
Moths	yes	no	yes	no	yes	no	yes	no	yes	no
Mice	yes	no	yes	no	yes	no	yes	no	yes	no
Ants	yes	no	yes	no	yes	no	yes	no	yes	no
Hive beetle	yes	no	yes	no	yes	no	yes	no	yes	no
Mold	yes	no	yes	no	yes	no	yes	no	yes	no
Varroa mites	yes	no	yes	no	yes	no	yes	no	yes	no
Other										
Overall Hive Health	1 2 3 4 5		1 2 3 4 5		1 2 3 4 5		1 2 3 4 5		1 2 3 4 5	

WHAT ACTIONS WERE TAKEN TODAY & WHAT IS TO BE SCHEDULED

HIVE INSPECTION

Date: _____ Time: _____ Temp: _____

Weather Conditions: _____

Inspector: _____

Location: _____

NAME: _____

GENERAL APPEARANCE	HIVE#		HIVE#		HIVE#		HIVE#		HIVE#	
Hive Temperment	☹	☺	☹	☺	☹	☺	☹	☺	☹	☺
Bees actively entering and exiting the hive?	yes	no	yes	no	yes	no	yes	no	yes	no
Are bees bringing pollen?	yes	no	yes	no	yes	no	yes	no	yes	no
Are there signs of disturbance?	yes	no	yes	no	yes	no	yes	no	yes	no
How many boxes?										

OPENING THE HIVES

Smoking material										
Capped honey	yes	no	yes	no	yes	no	yes	no	yes	no
Uncapped honey	yes	no	yes	no	yes	no	yes	no	yes	no
Capped brood	yes	no	yes	no	yes	no	yes	no	yes	no
Pollen	yes	no	yes	no	yes	no	yes	no	yes	no
Are there queen cells?	yes	no	yes	no	yes	no	yes	no	yes	no
Did you locate the queen?	yes	no	yes	no	yes	no	yes	no	yes	no
How many frames are covered in bees?										
Need space for nectar?	yes	no	yes	no	yes	no	yes	no	yes	no

PEST CONTROL Any signs of?

Bald brood	yes	no	yes	no	yes	no	yes	no	yes	no
Foul brood	yes	no	yes	no	yes	no	yes	no	yes	no
Bad smell	yes	no	yes	no	yes	no	yes	no	yes	no
Moths	yes	no	yes	no	yes	no	yes	no	yes	no
Mice	yes	no	yes	no	yes	no	yes	no	yes	no
Ants	yes	no	yes	no	yes	no	yes	no	yes	no
Hive beetle	yes	no	yes	no	yes	no	yes	no	yes	no
Mold	yes	no	yes	no	yes	no	yes	no	yes	no
Varroa mites	yes	no	yes	no	yes	no	yes	no	yes	no
Other										
Overall Hive Health	1 2 3 4 5		1 2 3 4 5		1 2 3 4 5		1 2 3 4 5		1 2 3 4 5	

WHAT ACTIONS WERE TAKEN TODAY & WHAT IS TO BE SCHEDULED

HIVE INSPECTION

Date: _____ Time: _____ Temp: _____

Weather Conditions: _____

Inspector: _____

Location: _____

NAME: _____

GENERAL APPEARANCE	HIVE#		HIVE#		HIVE#		HIVE#		HIVE#	
Hive Temperment	😠	😊	😠	😊	😠	😊	😠	😊	😠	😊
Bees actively entering and exiting the hive?	yes	no	yes	no	yes	no	yes	no	yes	no
Are bees bringing pollen?	yes	no	yes	no	yes	no	yes	no	yes	no
Are there signs of disturbance?	yes	no	yes	no	yes	no	yes	no	yes	no
How many boxes?										

OPENING THE HIVES

Smoking material										
Capped honey	yes	no	yes	no	yes	no	yes	no	yes	no
Uncapped honey	yes	no	yes	no	yes	no	yes	no	yes	no
Capped brood	yes	no	yes	no	yes	no	yes	no	yes	no
Pollen	yes	no	yes	no	yes	no	yes	no	yes	no
Are there queen cells?	yes	no	yes	no	yes	no	yes	no	yes	no
Did you locate the queen?	yes	no	yes	no	yes	no	yes	no	yes	no
How many frames are covered in bees?										
Need space for nectar?	yes	no	yes	no	yes	no	yes	no	yes	no

PEST CONTROL Any signs of?

Bald brood	yes	no	yes	no	yes	no	yes	no	yes	no
Foul brood	yes	no	yes	no	yes	no	yes	no	yes	no
Bad smell	yes	no	yes	no	yes	no	yes	no	yes	no
Moths	yes	no	yes	no	yes	no	yes	no	yes	no
Mice	yes	no	yes	no	yes	no	yes	no	yes	no
Ants	yes	no	yes	no	yes	no	yes	no	yes	no
Hive beetle	yes	no	yes	no	yes	no	yes	no	yes	no
Mold	yes	no	yes	no	yes	no	yes	no	yes	no
Varroa mites	yes	no	yes	no	yes	no	yes	no	yes	no
Other										
Overall Hive Health	1 2 3 4 5		1 2 3 4 5		1 2 3 4 5		1 2 3 4 5		1 2 3 4 5	

WHAT ACTIONS WERE TAKEN TODAY & WHAT IS TO BE SCHEDULED

_____

_____

_____

_____

_____

_____

_____

_____

HIVE INSPECTION

Date: _____ Time: _____ Temp: _____

Weather Conditions: _____

Inspector: _____ Location: _____

NAME: _____

GENERAL APPEARANCE	HIVE#		HIVE#		HIVE#		HIVE#		HIVE#	
Hive Temperment	😠	😊	😠	😊	😠	😊	😠	😊	😠	😊
Bees actively entering and exiting the hive?	yes	no	yes	no	yes	no	yes	no	yes	no
Are bees bringing pollen?	yes	no	yes	no	yes	no	yes	no	yes	no
Are there signs of disturbance?	yes	no	yes	no	yes	no	yes	no	yes	no
How many boxes?										

OPENING THE HIVES										
Smoking material										
Capped honey	yes	no	yes	no	yes	no	yes	no	yes	no
Uncapped honey	yes	no	yes	no	yes	no	yes	no	yes	no
Capped brood	yes	no	yes	no	yes	no	yes	no	yes	no
Pollen	yes	no	yes	no	yes	no	yes	no	yes	no
Are there queen cells?	yes	no	yes	no	yes	no	yes	no	yes	no
Did you locate the queen?	yes	no	yes	no	yes	no	yes	no	yes	no
How many frames are covered in bees?										
Need space for nectar?	yes	no	yes	no	yes	no	yes	no	yes	no

PEST CONTROL Any signs of?										
Bald brood	yes	no	yes	no	yes	no	yes	no	yes	no
Foul brood	yes	no	yes	no	yes	no	yes	no	yes	no
Bad smell	yes	no	yes	no	yes	no	yes	no	yes	no
Moths	yes	no	yes	no	yes	no	yes	no	yes	no
Mice	yes	no	yes	no	yes	no	yes	no	yes	no
Ants	yes	no	yes	no	yes	no	yes	no	yes	no
Hive beetle	yes	no	yes	no	yes	no	yes	no	yes	no
Mold	yes	no	yes	no	yes	no	yes	no	yes	no
Varroa mites	yes	no	yes	no	yes	no	yes	no	yes	no
Other										
Overall Hive Health	1 2 3 4 5		1 2 3 4 5		1 2 3 4 5		1 2 3 4 5		1 2 3 4 5	

WHAT ACTIONS WERE TAKEN TODAY & WHAT IS TO BE SCHEDULED

HIVE INSPECTION

Date: _____ Time: _____ Temp: _____

Weather Conditions: _____

Inspector: _____ Location: _____

NAME: _____

GENERAL APPEARANCE	HIVE#		HIVE#		HIVE#		HIVE#		HIVE#	
Hive Temperment	☹	☺	☹	☺	☹	☺	☹	☺	☹	☺
Bees actively entering and exiting the hive?	yes	no	yes	no	yes	no	yes	no	yes	no
Are bees bringing pollen?	yes	no	yes	no	yes	no	yes	no	yes	no
Are there signs of disturbance?	yes	no	yes	no	yes	no	yes	no	yes	no
How many boxes?										

OPENING THE HIVES										
Smoking material										
Capped honey	yes	no	yes	no	yes	no	yes	no	yes	no
Uncapped honey	yes	no	yes	no	yes	no	yes	no	yes	no
Capped brood	yes	no	yes	no	yes	no	yes	no	yes	no
Pollen	yes	no	yes	no	yes	no	yes	no	yes	no
Are there queen cells?	yes	no	yes	no	yes	no	yes	no	yes	no
Did you locate the queen?	yes	no	yes	no	yes	no	yes	no	yes	no
How many frames are covered in bees?										
Need space for nectar?	yes	no	yes	no	yes	no	yes	no	yes	no

PEST CONTROL Any signs of?										
Bald brood	yes	no	yes	no	yes	no	yes	no	yes	no
Foul brood	yes	no	yes	no	yes	no	yes	no	yes	no
Bad smell	yes	no	yes	no	yes	no	yes	no	yes	no
Moths	yes	no	yes	no	yes	no	yes	no	yes	no
Mice	yes	no	yes	no	yes	no	yes	no	yes	no
Ants	yes	no	yes	no	yes	no	yes	no	yes	no
Hive beetle	yes	no	yes	no	yes	no	yes	no	yes	no
Mold	yes	no	yes	no	yes	no	yes	no	yes	no
Varroa mites	yes	no	yes	no	yes	no	yes	no	yes	no
Other										
Overall Hive Health	1 2 3 4 5		1 2 3 4 5		1 2 3 4 5		1 2 3 4 5		1 2 3 4 5	

WHAT ACTIONS WERE TAKEN TODAY & WHAT IS TO BE SCHEDULED

HIVE INSPECTION

Date: _____ Time: _____ Temp: _____

Weather Conditions: _____

Inspector: _____

Location: _____

NAME: _____

GENERAL APPEARANCE	HIVE#		HIVE#		HIVE#		HIVE#		HIVE#	
Hive Temperment	😠	😊	😠	😊	😠	😊	😠	😊	😠	😊
Bees actively entering and exiting the hive?	yes	no	yes	no	yes	no	yes	no	yes	no
Are bees bringing pollen?	yes	no	yes	no	yes	no	yes	no	yes	no
Are there signs of disturbance?	yes	no	yes	no	yes	no	yes	no	yes	no
How many boxes?										

OPENING THE HIVES										
Smoking material										
Capped honey	yes	no	yes	no	yes	no	yes	no	yes	no
Uncapped honey	yes	no	yes	no	yes	no	yes	no	yes	no
Capped brood	yes	no	yes	no	yes	no	yes	no	yes	no
Pollen	yes	no	yes	no	yes	no	yes	no	yes	no
Are there queen cells?	yes	no	yes	no	yes	no	yes	no	yes	no
Did you locate the queen?	yes	no	yes	no	yes	no	yes	no	yes	no
How many frames are covered in bees?										
Need space for nectar?	yes	no	yes	no	yes	no	yes	no	yes	no

PEST CONTROL Any signs of?

Bald brood	yes	no	yes	no	yes	no	yes	no	yes	no
Foul brood	yes	no	yes	no	yes	no	yes	no	yes	no
Bad smell	yes	no	yes	no	yes	no	yes	no	yes	no
Moths	yes	no	yes	no	yes	no	yes	no	yes	no
Mice	yes	no	yes	no	yes	no	yes	no	yes	no
Ants	yes	no	yes	no	yes	no	yes	no	yes	no
Hive beetle	yes	no	yes	no	yes	no	yes	no	yes	no
Mold	yes	no	yes	no	yes	no	yes	no	yes	no
Varroa mites	yes	no	yes	no	yes	no	yes	no	yes	no
Other										
Overall Hive Health	1 2 3 4 5		1 2 3 4 5		1 2 3 4 5		1 2 3 4 5		1 2 3 4 5	

WHAT ACTIONS WERE TAKEN TODAY & WHAT IS TO BE SCHEDULED

HIVE INSPECTION

Date: _____ Time: _____ Temp: _____

Weather Conditions: _____

Inspector: _____ Location: _____

NAME: _____

GENERAL APPEARANCE	HIVE#		HIVE#		HIVE#		HIVE#		HIVE#	
Hive Temperment	☹	☺	☹	☺	☹	☺	☹	☺	☹	☺
Bees actively entering and exiting the hive?	yes	no	yes	no	yes	no	yes	no	yes	no
Are bees bringing pollen?	yes	no	yes	no	yes	no	yes	no	yes	no
Are there signs of disturbance?	yes	no	yes	no	yes	no	yes	no	yes	no
How many boxes?										

OPENING THE HIVES										
Smoking material										
Capped honey	yes	no	yes	no	yes	no	yes	no	yes	no
Uncapped honey	yes	no	yes	no	yes	no	yes	no	yes	no
Capped brood	yes	no	yes	no	yes	no	yes	no	yes	no
Pollen	yes	no	yes	no	yes	no	yes	no	yes	no
Are there queen cells?	yes	no	yes	no	yes	no	yes	no	yes	no
Did you locate the queen?	yes	no	yes	no	yes	no	yes	no	yes	no
How many frames are covered in bees?										
Need space for nectar?	yes	no	yes	no	yes	no	yes	no	yes	no

PEST CONTROL Any signs of?										
Bald brood	yes	no	yes	no	yes	no	yes	no	yes	no
Foul brood	yes	no	yes	no	yes	no	yes	no	yes	no
Bad smell	yes	no	yes	no	yes	no	yes	no	yes	no
Moths	yes	no	yes	no	yes	no	yes	no	yes	no
Mice	yes	no	yes	no	yes	no	yes	no	yes	no
Ants	yes	no	yes	no	yes	no	yes	no	yes	no
Hive beetle	yes	no	yes	no	yes	no	yes	no	yes	no
Mold	yes	no	yes	no	yes	no	yes	no	yes	no
Varroa mites	yes	no	yes	no	yes	no	yes	no	yes	no
Other										
Overall Hive Health	1 2 3 4 5		1 2 3 4 5		1 2 3 4 5		1 2 3 4 5		1 2 3 4 5	

WHAT ACTIONS WERE TAKEN TODAY & WHAT IS TO BE SCHEDULED

_____

_____

_____

_____

_____

_____

_____

_____

HIVE INSPECTION

Date: _____ Time: _____ Temp: _____

Weather Conditions: _____

Inspector: _____ Location: _____

NAME:

GENERAL APPEARANCE	HIVE#		HIVE#		HIVE#		HIVE#		HIVE#	
Hive Temperment	☹	☺	☹	☺	☹	☺	☹	☺	☹	☺
Bees actively entering and exiting the hive?	yes	no	yes	no	yes	no	yes	no	yes	no
Are bees bringing pollen?	yes	no	yes	no	yes	no	yes	no	yes	no
Are there signs of disturbance?	yes	no	yes	no	yes	no	yes	no	yes	no
How many boxes?										

OPENING THE HIVES

Smoking material										
Capped honey	yes	no	yes	no	yes	no	yes	no	yes	no
Uncapped honey	yes	no	yes	no	yes	no	yes	no	yes	no
Capped brood	yes	no	yes	no	yes	no	yes	no	yes	no
Pollen	yes	no	yes	no	yes	no	yes	no	yes	no
Are there queen cells?	yes	no	yes	no	yes	no	yes	no	yes	no
Did you locate the queen?	yes	no	yes	no	yes	no	yes	no	yes	no
How many frames are covered in bees?										
Need space for nectar?	yes	no	yes	no	yes	no	yes	no	yes	no

PEST CONTROL Any signs of?

Bald brood	yes	no	yes	no	yes	no	yes	no	yes	no
Foul brood	yes	no	yes	no	yes	no	yes	no	yes	no
Bad smell	yes	no	yes	no	yes	no	yes	no	yes	no
Moths	yes	no	yes	no	yes	no	yes	no	yes	no
Mice	yes	no	yes	no	yes	no	yes	no	yes	no
Ants	yes	no	yes	no	yes	no	yes	no	yes	no
Hive beetle	yes	no	yes	no	yes	no	yes	no	yes	no
Mold	yes	no	yes	no	yes	no	yes	no	yes	no
Varroa mites	yes	no	yes	no	yes	no	yes	no	yes	no
Other										
Overall Hive Health	1 2 3 4 5		1 2 3 4 5		1 2 3 4 5		1 2 3 4 5		1 2 3 4 5	

WHAT ACTIONS WERE TAKEN TODAY & WHAT IS TO BE SCHEDULED

HIVE INSPECTION

Date: _____ Time: _____ Temp: _____

Weather Conditions: _____

Inspector: _____

Location: _____

NAME: _____

GENERAL APPEARANCE	HIVE#		HIVE#		HIVE#		HIVE#		HIVE#	
Hive Temperment	☹	☺	☹	☺	☹	☺	☹	☺	☹	☺
Bees actively entering and exiting the hive?	yes	no	yes	no	yes	no	yes	no	yes	no
Are bees bringing pollen?	yes	no	yes	no	yes	no	yes	no	yes	no
Are there signs of disturbance?	yes	no	yes	no	yes	no	yes	no	yes	no
How many boxes?										

OPENING THE HIVES

Smoking material										
Capped honey	yes	no	yes	no	yes	no	yes	no	yes	no
Uncapped honey	yes	no	yes	no	yes	no	yes	no	yes	no
Capped brood	yes	no	yes	no	yes	no	yes	no	yes	no
Pollen	yes	no	yes	no	yes	no	yes	no	yes	no
Are there queen cells?	yes	no	yes	no	yes	no	yes	no	yes	no
Did you locate the queen?	yes	no	yes	no	yes	no	yes	no	yes	no
How many frames are covered in bees?										
Need space for nectar?	yes	no	yes	no	yes	no	yes	no	yes	no

PEST CONTROL Any signs of?

Bald brood	yes	no	yes	no	yes	no	yes	no	yes	no
Foul brood	yes	no	yes	no	yes	no	yes	no	yes	no
Bad smell	yes	no	yes	no	yes	no	yes	no	yes	no
Moths	yes	no	yes	no	yes	no	yes	no	yes	no
Mice	yes	no	yes	no	yes	no	yes	no	yes	no
Ants	yes	no	yes	no	yes	no	yes	no	yes	no
Hive beetle	yes	no	yes	no	yes	no	yes	no	yes	no
Mold	yes	no	yes	no	yes	no	yes	no	yes	no
Varroa mites	yes	no	yes	no	yes	no	yes	no	yes	no
Other										
Overall Hive Health	1 2 3 4 5		1 2 3 4 5		1 2 3 4 5		1 2 3 4 5		1 2 3 4 5	

WHAT ACTIONS WERE TAKEN TODAY & WHAT IS TO BE SCHEDULED

HIVE INSPECTION

Date: _____ Time: _____ Temp: _____

Weather Conditions: _____

Inspector: _____   Location: _____

NAME: _____

GENERAL APPEARANCE	HIVE#		HIVE#		HIVE#		HIVE#		HIVE#	
Hive Temperment	☹	☺	☹	☺	☹	☺	☹	☺	☹	☺
Bees actively entering and exiting the hive?	yes	no	yes	no	yes	no	yes	no	yes	no
Are bees bringing pollen?	yes	no	yes	no	yes	no	yes	no	yes	no
Are there signs of disturbance?	yes	no	yes	no	yes	no	yes	no	yes	no
How many boxes?										

OPENING THE HIVES

Smoking material										
Capped honey	yes	no	yes	no	yes	no	yes	no	yes	no
Uncapped honey	yes	no	yes	no	yes	no	yes	no	yes	no
Capped brood	yes	no	yes	no	yes	no	yes	no	yes	no
Pollen	yes	no	yes	no	yes	no	yes	no	yes	no
Are there queen cells?	yes	no	yes	no	yes	no	yes	no	yes	no
Did you locate the queen?	yes	no	yes	no	yes	no	yes	no	yes	no
How many frames are covered in bees?										
Need space for nectar?	yes	no	yes	no	yes	no	yes	no	yes	no

PEST CONTROL Any signs of?

Bald brood	yes	no	yes	no	yes	no	yes	no	yes	no
Foul brood	yes	no	yes	no	yes	no	yes	no	yes	no
Bad smell	yes	no	yes	no	yes	no	yes	no	yes	no
Moths	yes	no	yes	no	yes	no	yes	no	yes	no
Mice	yes	no	yes	no	yes	no	yes	no	yes	no
Ants	yes	no	yes	no	yes	no	yes	no	yes	no
Hive beetle	yes	no	yes	no	yes	no	yes	no	yes	no
Mold	yes	no	yes	no	yes	no	yes	no	yes	no
Varroa mites	yes	no	yes	no	yes	no	yes	no	yes	no
Other										
Overall Hive Health	1 2 3 4 5		1 2 3 4 5		1 2 3 4 5		1 2 3 4 5		1 2 3 4 5	

WHAT ACTIONS WERE TAKEN TODAY & WHAT IS TO BE SCHEDULED

HIVE INSPECTION

Date: _____ Time: _____ Temp: _____

Weather Conditions: _____

Inspector: _____

Location: _____

NAME: _____

GENERAL APPEARANCE	HIVE#		HIVE#		HIVE#		HIVE#		HIVE#	
Hive Temperment	☹	☺	☹	☺	☹	☺	☹	☺	☹	☺
Bees actively entering and exiting the hive?	yes	no	yes	no	yes	no	yes	no	yes	no
Are bees bringing pollen?	yes	no	yes	no	yes	no	yes	no	yes	no
Are there signs of disturbance?	yes	no	yes	no	yes	no	yes	no	yes	no
How many boxes?										

OPENING THE HIVES										
Smoking material										
Capped honey	yes	no	yes	no	yes	no	yes	no	yes	no
Uncapped honey	yes	no	yes	no	yes	no	yes	no	yes	no
Capped brood	yes	no	yes	no	yes	no	yes	no	yes	no
Pollen	yes	no	yes	no	yes	no	yes	no	yes	no
Are there queen cells?	yes	no	yes	no	yes	no	yes	no	yes	no
Did you locate the queen?	yes	no	yes	no	yes	no	yes	no	yes	no
How many frames are covered in bees?										
Need space for nectar?	yes	no	yes	no	yes	no	yes	no	yes	no

PEST CONTROL Any signs of?										
Bald brood	yes	no	yes	no	yes	no	yes	no	yes	no
Foul brood	yes	no	yes	no	yes	no	yes	no	yes	no
Bad smell	yes	no	yes	no	yes	no	yes	no	yes	no
Moths	yes	no	yes	no	yes	no	yes	no	yes	no
Mice	yes	no	yes	no	yes	no	yes	no	yes	no
Ants	yes	no	yes	no	yes	no	yes	no	yes	no
Hive beetle	yes	no	yes	no	yes	no	yes	no	yes	no
Mold	yes	no	yes	no	yes	no	yes	no	yes	no
Varroa mites	yes	no	yes	no	yes	no	yes	no	yes	no
Other										
Overall Hive Health	1 2 3 4 5		1 2 3 4 5		1 2 3 4 5		1 2 3 4 5		1 2 3 4 5	

WHAT ACTIONS WERE TAKEN TODAY & WHAT IS TO BE SCHEDULED

HIVE INSPECTION

Date: _____   Time: _____   Temp: _____

Weather Conditions: _____

Inspector: _____   Location: _____

NAME:										
GENERAL APPEARANCE	HIVE#		HIVE#		HIVE#		HIVE#		HIVE#	
Hive Temperment	☹	☺	☹	☺	☹	☺	☹	☺	☹	☺
Bees actively entering and exiting the hive?	yes	no	yes	no	yes	no	yes	no	yes	no
Are bees bringing pollen?	yes	no	yes	no	yes	no	yes	no	yes	no
Are there signs of disturbance?	yes	no	yes	no	yes	no	yes	no	yes	no
How many boxes?										
OPENING THE HIVES										
Smoking material										
Capped honey	yes	no	yes	no	yes	no	yes	no	yes	no
Uncapped honey	yes	no	yes	no	yes	no	yes	no	yes	no
Capped brood	yes	no	yes	no	yes	no	yes	no	yes	no
Pollen	yes	no	yes	no	yes	no	yes	no	yes	no
Are there queen cells?	yes	no	yes	no	yes	no	yes	no	yes	no
Did you locate the queen?	yes	no	yes	no	yes	no	yes	no	yes	no
How many frames are covered in bees?										
Need space for nectar?	yes	no	yes	no	yes	no	yes	no	yes	no
PEST CONTROL Any signs of?										
Bald brood	yes	no	yes	no	yes	no	yes	no	yes	no
Foul brood	yes	no	yes	no	yes	no	yes	no	yes	no
Bad smell	yes	no	yes	no	yes	no	yes	no	yes	no
Moths	yes	no	yes	no	yes	no	yes	no	yes	no
Mice	yes	no	yes	no	yes	no	yes	no	yes	no
Ants	yes	no	yes	no	yes	no	yes	no	yes	no
Hive beetle	yes	no	yes	no	yes	no	yes	no	yes	no
Mold	yes	no	yes	no	yes	no	yes	no	yes	no
Varroa mites	yes	no	yes	no	yes	no	yes	no	yes	no
Other										
Overall Hive Health	1 2 3 4 5		1 2 3 4 5		1 2 3 4 5		1 2 3 4 5		1 2 3 4 5	

WHAT ACTIONS WERE TAKEN TODAY & WHAT IS TO BE SCHEDULED

HIVE INSPECTION

Date: _____ Time: _____ Temp: _____

Weather Conditions: _____

Inspector: _____

Location: _____

NAME: _____

GENERAL APPEARANCE	HIVE#		HIVE#		HIVE#		HIVE#		HIVE#	
Hive Temperment	😠	😊	😠	😊	😠	😊	😠	😊	😠	😊
Bees actively entering and exiting the hive?	yes	no	yes	no	yes	no	yes	no	yes	no
Are bees bringing pollen?	yes	no	yes	no	yes	no	yes	no	yes	no
Are there signs of disturbance?	yes	no	yes	no	yes	no	yes	no	yes	no
How many boxes?										
OPENING THE HIVES										
Smoking material										
Capped honey	yes	no	yes	no	yes	no	yes	no	yes	no
Uncapped honey	yes	no	yes	no	yes	no	yes	no	yes	no
Capped brood	yes	no	yes	no	yes	no	yes	no	yes	no
Pollen	yes	no	yes	no	yes	no	yes	no	yes	no
Are there queen cells?	yes	no	yes	no	yes	no	yes	no	yes	no
Did you locate the queen?	yes	no	yes	no	yes	no	yes	no	yes	no
How many frames are covered in bees?										
Need space for nectar?	yes	no	yes	no	yes	no	yes	no	yes	no
PEST CONTROL Any signs of?										
Bald brood	yes	no	yes	no	yes	no	yes	no	yes	no
Foul brood	yes	no	yes	no	yes	no	yes	no	yes	no
Bad smell	yes	no	yes	no	yes	no	yes	no	yes	no
Moths	yes	no	yes	no	yes	no	yes	no	yes	no
Mice	yes	no	yes	no	yes	no	yes	no	yes	no
Ants	yes	no	yes	no	yes	no	yes	no	yes	no
Hive beetle	yes	no	yes	no	yes	no	yes	no	yes	no
Mold	yes	no	yes	no	yes	no	yes	no	yes	no
Varroa mites	yes	no	yes	no	yes	no	yes	no	yes	no
Other										
Overall Hive Health	1 2 3 4 5		1 2 3 4 5		1 2 3 4 5		1 2 3 4 5		1 2 3 4 5	

WHAT ACTIONS WERE TAKEN TODAY & WHAT IS TO BE SCHEDULED

HIVE INSPECTION

Date: _____ Time: _____ Temp: _____

Weather Conditions: _____

Inspector: _____

Location: _____

NAME: _____

GENERAL APPEARANCE	HIVE#		HIVE#		HIVE#		HIVE#		HIVE#	
Hive Temperment	😖	😊	😖	😊	😖	😊	😖	😊	😖	😊
Bees actively entering and exiting the hive?	yes	no	yes	no	yes	no	yes	no	yes	no
Are bees bringing pollen?	yes	no	yes	no	yes	no	yes	no	yes	no
Are there signs of disturbance?	yes	no	yes	no	yes	no	yes	no	yes	no
How many boxes?										

OPENING THE HIVES										
Smoking material										
Capped honey	yes	no	yes	no	yes	no	yes	no	yes	no
Uncapped honey	yes	no	yes	no	yes	no	yes	no	yes	no
Capped brood	yes	no	yes	no	yes	no	yes	no	yes	no
Pollen	yes	no	yes	no	yes	no	yes	no	yes	no
Are there queen cells?	yes	no	yes	no	yes	no	yes	no	yes	no
Did you locate the queen?	yes	no	yes	no	yes	no	yes	no	yes	no
How many frames are covered in bees?										
Need space for nectar?	yes	no	yes	no	yes	no	yes	no	yes	no

PEST CONTROL Any signs of?										
Bald brood	yes	no	yes	no	yes	no	yes	no	yes	no
Foul brood	yes	no	yes	no	yes	no	yes	no	yes	no
Bad smell	yes	no	yes	no	yes	no	yes	no	yes	no
Moths	yes	no	yes	no	yes	no	yes	no	yes	no
Mice	yes	no	yes	no	yes	no	yes	no	yes	no
Ants	yes	no	yes	no	yes	no	yes	no	yes	no
Hive beetle	yes	no	yes	no	yes	no	yes	no	yes	no
Mold	yes	no	yes	no	yes	no	yes	no	yes	no
Varroa mites	yes	no	yes	no	yes	no	yes	no	yes	no
Other										
Overall Hive Health	1 2 3 4 5		1 2 3 4 5		1 2 3 4 5		1 2 3 4 5		1 2 3 4 5	

WHAT ACTIONS WERE TAKEN TODAY & WHAT IS TO BE SCHEDULED

HIVE INSPECTION

Date: _____ Time: _____ Temp: _____

Weather Conditions: _____

Inspector: _____

Location: _____

NAME: _____

GENERAL APPEARANCE	HIVE#		HIVE#		HIVE#		HIVE#		HIVE#	
Hive Temperment	😠	😊	😠	😊	😠	😊	😠	😊	😠	😊
Bees actively entering and exiting the hive?	yes	no	yes	no	yes	no	yes	no	yes	no
Are bees bringing pollen?	yes	no	yes	no	yes	no	yes	no	yes	no
Are there signs of disturbance?	yes	no	yes	no	yes	no	yes	no	yes	no
How many boxes?										

OPENING THE HIVES										
Smoking material										
Capped honey	yes	no	yes	no	yes	no	yes	no	yes	no
Uncapped honey	yes	no	yes	no	yes	no	yes	no	yes	no
Capped brood	yes	no	yes	no	yes	no	yes	no	yes	no
Pollen	yes	no	yes	no	yes	no	yes	no	yes	no
Are there queen cells?	yes	no	yes	no	yes	no	yes	no	yes	no
Did you locate the queen?	yes	no	yes	no	yes	no	yes	no	yes	no
How many frames are covered in bees?										
Need space for nectar?	yes	no	yes	no	yes	no	yes	no	yes	no

PEST CONTROL Any signs of?										
Bald brood	yes	no	yes	no	yes	no	yes	no	yes	no
Foul brood	yes	no	yes	no	yes	no	yes	no	yes	no
Bad smell	yes	no	yes	no	yes	no	yes	no	yes	no
Moths	yes	no	yes	no	yes	no	yes	no	yes	no
Mice	yes	no	yes	no	yes	no	yes	no	yes	no
Ants	yes	no	yes	no	yes	no	yes	no	yes	no
Hive beetle	yes	no	yes	no	yes	no	yes	no	yes	no
Mold	yes	no	yes	no	yes	no	yes	no	yes	no
Varroa mites	yes	no	yes	no	yes	no	yes	no	yes	no
Other										
Overall Hive Health	1 2 3 4 5		1 2 3 4 5		1 2 3 4 5		1 2 3 4 5		1 2 3 4 5	

WHAT ACTIONS WERE TAKEN TODAY & WHAT IS TO BE SCHEDULED

_____

_____

_____

_____

_____

_____

_____

HIVE INSPECTION

Date: _____ Time: _____ Temp: _____

Weather Conditions: _____

Inspector: _____ Location: _____

NAME: _____

GENERAL APPEARANCE	HIVE#		HIVE#		HIVE#		HIVE#		HIVE#	
Hive Temperment	☹	☺	☹	☺	☹	☺	☹	☺	☹	☺
Bees actively entering and exiting the hive?	yes	no	yes	no	yes	no	yes	no	yes	no
Are bees bringing pollen?	yes	no	yes	no	yes	no	yes	no	yes	no
Are there signs of disturbance?	yes	no	yes	no	yes	no	yes	no	yes	no
How many boxes?										

OPENING THE HIVES										
Smoking material										
Capped honey	yes	no	yes	no	yes	no	yes	no	yes	no
Uncapped honey	yes	no	yes	no	yes	no	yes	no	yes	no
Capped brood	yes	no	yes	no	yes	no	yes	no	yes	no
Pollen	yes	no	yes	no	yes	no	yes	no	yes	no
Are there queen cells?	yes	no	yes	no	yes	no	yes	no	yes	no
Did you locate the queen?	yes	no	yes	no	yes	no	yes	no	yes	no
How many frames are covered in bees?										
Need space for nectar?	yes	no	yes	no	yes	no	yes	no	yes	no

PEST CONTROL Any signs of?										
Bald brood	yes	no	yes	no	yes	no	yes	no	yes	no
Foul brood	yes	no	yes	no	yes	no	yes	no	yes	no
Bad smell	yes	no	yes	no	yes	no	yes	no	yes	no
Moths	yes	no	yes	no	yes	no	yes	no	yes	no
Mice	yes	no	yes	no	yes	no	yes	no	yes	no
Ants	yes	no	yes	no	yes	no	yes	no	yes	no
Hive beetle	yes	no	yes	no	yes	no	yes	no	yes	no
Mold	yes	no	yes	no	yes	no	yes	no	yes	no
Varroa mites	yes	no	yes	no	yes	no	yes	no	yes	no
Other										
Overall Hive Health	1 2 3 4 5		1 2 3 4 5		1 2 3 4 5		1 2 3 4 5		1 2 3 4 5	

WHAT ACTIONS WERE TAKEN TODAY & WHAT IS TO BE SCHEDULED

HIVE INSPECTION

Date: _____ Time: _____ Temp: _____

Weather Conditions: _____

Inspector: _____ Location: _____

NAME: _____

GENERAL APPEARANCE	HIVE#		HIVE#		HIVE#		HIVE#		HIVE#	
Hive Temperment	☹	☺	☹	☺	☹	☺	☹	☺	☹	☺
Bees actively entering and exiting the hive?	yes	no	yes	no	yes	no	yes	no	yes	no
Are bees bringing pollen?	yes	no	yes	no	yes	no	yes	no	yes	no
Are there signs of disturbance?	yes	no	yes	no	yes	no	yes	no	yes	no
How many boxes?										

OPENING THE HIVES										
Smoking material										
Capped honey	yes	no	yes	no	yes	no	yes	no	yes	no
Uncapped honey	yes	no	yes	no	yes	no	yes	no	yes	no
Capped brood	yes	no	yes	no	yes	no	yes	no	yes	no
Pollen	yes	no	yes	no	yes	no	yes	no	yes	no
Are there queen cells?	yes	no	yes	no	yes	no	yes	no	yes	no
Did you locate the queen?	yes	no	yes	no	yes	no	yes	no	yes	no
How many frames are covered in bees?										
Need space for nectar?	yes	no	yes	no	yes	no	yes	no	yes	no

PEST CONTROL Any signs of?										
Bald brood	yes	no	yes	no	yes	no	yes	no	yes	no
Foul brood	yes	no	yes	no	yes	no	yes	no	yes	no
Bad smell	yes	no	yes	no	yes	no	yes	no	yes	no
Moths	yes	no	yes	no	yes	no	yes	no	yes	no
Mice	yes	no	yes	no	yes	no	yes	no	yes	no
Ants	yes	no	yes	no	yes	no	yes	no	yes	no
Hive beetle	yes	no	yes	no	yes	no	yes	no	yes	no
Mold	yes	no	yes	no	yes	no	yes	no	yes	no
Varroa mites	yes	no	yes	no	yes	no	yes	no	yes	no
Other										
Overall Hive Health	1 2 3 4 5		1 2 3 4 5		1 2 3 4 5		1 2 3 4 5		1 2 3 4 5	

WHAT ACTIONS WERE TAKEN TODAY & WHAT IS TO BE SCHEDULED

HIVE INSPECTION

Date: _____ Time: _____ Temp: _____

Weather Conditions: _____

Inspector: _____ Location: _____

NAME: _____

GENERAL APPEARANCE	HIVE#		HIVE#		HIVE#		HIVE#		HIVE#	
Hive Temperment	☹	☺	☹	☺	☹	☺	☹	☺	☹	☺
Bees actively entering and exiting the hive?	yes	no	yes	no	yes	no	yes	no	yes	no
Are bees bringing pollen?	yes	no	yes	no	yes	no	yes	no	yes	no
Are there signs of disturbance?	yes	no	yes	no	yes	no	yes	no	yes	no
How many boxes?										

OPENING THE HIVES										
Smoking material										
Capped honey	yes	no	yes	no	yes	no	yes	no	yes	no
Uncapped honey	yes	no	yes	no	yes	no	yes	no	yes	no
Capped brood	yes	no	yes	no	yes	no	yes	no	yes	no
Pollen	yes	no	yes	no	yes	no	yes	no	yes	no
Are there queen cells?	yes	no	yes	no	yes	no	yes	no	yes	no
Did you locate the queen?	yes	no	yes	no	yes	no	yes	no	yes	no
How many frames are covered in bees?										
Need space for nectar?	yes	no	yes	no	yes	no	yes	no	yes	no

PEST CONTROL Any signs of?										
Bald brood	yes	no	yes	no	yes	no	yes	no	yes	no
Foul brood	yes	no	yes	no	yes	no	yes	no	yes	no
Bad smell	yes	no	yes	no	yes	no	yes	no	yes	no
Moths	yes	no	yes	no	yes	no	yes	no	yes	no
Mice	yes	no	yes	no	yes	no	yes	no	yes	no
Ants	yes	no	yes	no	yes	no	yes	no	yes	no
Hive beetle	yes	no	yes	no	yes	no	yes	no	yes	no
Mold	yes	no	yes	no	yes	no	yes	no	yes	no
Varroa mites	yes	no	yes	no	yes	no	yes	no	yes	no
Other										
Overall Hive Health	1 2 3 4 5		1 2 3 4 5		1 2 3 4 5		1 2 3 4 5		1 2 3 4 5	

WHAT ACTIONS WERE TAKEN TODAY & WHAT IS TO BE SCHEDULED

_____

_____

_____

_____

_____

_____

_____

_____

_____

_____

HIVE INSPECTION

Date: _____ Time: _____ Temp: _____

Weather Conditions: _____

Inspector: _____

Location: _____

NAME: _____

GENERAL APPEARANCE	HIVE#		HIVE#		HIVE#		HIVE#		HIVE#	
Hive Temperment	☹	☺	☹	☺	☹	☺	☹	☺	☹	☺
Bees actively entering and exiting the hive?	yes	no	yes	no	yes	no	yes	no	yes	no
Are bees bringing pollen?	yes	no	yes	no	yes	no	yes	no	yes	no
Are there signs of disturbance?	yes	no	yes	no	yes	no	yes	no	yes	no
How many boxes?										

OPENING THE HIVES										
Smoking material										
Capped honey	yes	no	yes	no	yes	no	yes	no	yes	no
Uncapped honey	yes	no	yes	no	yes	no	yes	no	yes	no
Capped brood	yes	no	yes	no	yes	no	yes	no	yes	no
Pollen	yes	no	yes	no	yes	no	yes	no	yes	no
Are there queen cells?	yes	no	yes	no	yes	no	yes	no	yes	no
Did you locate the queen?	yes	no	yes	no	yes	no	yes	no	yes	no
How many frames are covered in bees?										
Need space for nectar?	yes	no	yes	no	yes	no	yes	no	yes	no

PEST CONTROL Any signs of?										
Bald brood	yes	no	yes	no	yes	no	yes	no	yes	no
Foul brood	yes	no	yes	no	yes	no	yes	no	yes	no
Bad smell	yes	no	yes	no	yes	no	yes	no	yes	no
Moths	yes	no	yes	no	yes	no	yes	no	yes	no
Mice	yes	no	yes	no	yes	no	yes	no	yes	no
Ants	yes	no	yes	no	yes	no	yes	no	yes	no
Hive beetle	yes	no	yes	no	yes	no	yes	no	yes	no
Mold	yes	no	yes	no	yes	no	yes	no	yes	no
Varroa mites	yes	no	yes	no	yes	no	yes	no	yes	no
Other										
Overall Hive Health	1 2 3 4 5		1 2 3 4 5		1 2 3 4 5		1 2 3 4 5		1 2 3 4 5	

WHAT ACTIONS WERE TAKEN TODAY & WHAT IS TO BE SCHEDULED

HIVE INSPECTION

Date: _____ Time: _____ Temp: _____

Weather Conditions: _____

Inspector: _____ Location: _____

NAME: _____

GENERAL APPEARANCE	HIVE#		HIVE#		HIVE#		HIVE#		HIVE#	
Hive Temperment	☹	☺	☹	☺	☹	☺	☹	☺	☹	☺
Bees actively entering and exiting the hive?	yes	no	yes	no	yes	no	yes	no	yes	no
Are bees bringing pollen?	yes	no	yes	no	yes	no	yes	no	yes	no
Are there signs of disturbance?	yes	no	yes	no	yes	no	yes	no	yes	no
How many boxes?										

OPENING THE HIVES										
Smoking material										
Capped honey	yes	no	yes	no	yes	no	yes	no	yes	no
Uncapped honey	yes	no	yes	no	yes	no	yes	no	yes	no
Capped brood	yes	no	yes	no	yes	no	yes	no	yes	no
Pollen	yes	no	yes	no	yes	no	yes	no	yes	no
Are there queen cells?	yes	no	yes	no	yes	no	yes	no	yes	no
Did you locate the queen?	yes	no	yes	no	yes	no	yes	no	yes	no
How many frames are covered in bees?										
Need space for nectar?	yes	no	yes	no	yes	no	yes	no	yes	no

PEST CONTROL Any signs of?										
Bald brood	yes	no	yes	no	yes	no	yes	no	yes	no
Foul brood	yes	no	yes	no	yes	no	yes	no	yes	no
Bad smell	yes	no	yes	no	yes	no	yes	no	yes	no
Moths	yes	no	yes	no	yes	no	yes	no	yes	no
Mice	yes	no	yes	no	yes	no	yes	no	yes	no
Ants	yes	no	yes	no	yes	no	yes	no	yes	no
Hive beetle	yes	no	yes	no	yes	no	yes	no	yes	no
Mold	yes	no	yes	no	yes	no	yes	no	yes	no
Varroa mites	yes	no	yes	no	yes	no	yes	no	yes	no
Other										
Overall Hive Health	1 2 3 4 5		1 2 3 4 5		1 2 3 4 5		1 2 3 4 5		1 2 3 4 5	

WHAT ACTIONS WERE TAKEN TODAY & WHAT IS TO BE SCHEDULED

HIVE INSPECTION

Date: _____ Time: _____ Temp: _____

Weather Conditions: _____

Inspector: _____

Location: _____

NAME: _____

GENERAL APPEARANCE	HIVE#		HIVE#		HIVE#		HIVE#		HIVE#	
Hive Temperment	☹	☺	☹	☺	☹	☺	☹	☺	☹	☺
Bees actively entering and exiting the hive?	yes	no	yes	no	yes	no	yes	no	yes	no
Are bees bringing pollen?	yes	no	yes	no	yes	no	yes	no	yes	no
Are there signs of disturbance?	yes	no	yes	no	yes	no	yes	no	yes	no
How many boxes?										
OPENING THE HIVES										
Smoking material										
Capped honey	yes	no	yes	no	yes	no	yes	no	yes	no
Uncapped honey	yes	no	yes	no	yes	no	yes	no	yes	no
Capped brood	yes	no	yes	no	yes	no	yes	no	yes	no
Pollen	yes	no	yes	no	yes	no	yes	no	yes	no
Are there queen cells?	yes	no	yes	no	yes	no	yes	no	yes	no
Did you locate the queen?	yes	no	yes	no	yes	no	yes	no	yes	no
How many frames are covered in bees?										
Need space for nectar?	yes	no	yes	no	yes	no	yes	no	yes	no
PEST CONTROL Any signs of?										
Bald brood	yes	no	yes	no	yes	no	yes	no	yes	no
Foul brood	yes	no	yes	no	yes	no	yes	no	yes	no
Bad smell	yes	no	yes	no	yes	no	yes	no	yes	no
Moths	yes	no	yes	no	yes	no	yes	no	yes	no
Mice	yes	no	yes	no	yes	no	yes	no	yes	no
Ants	yes	no	yes	no	yes	no	yes	no	yes	no
Hive beetle	yes	no	yes	no	yes	no	yes	no	yes	no
Mold	yes	no	yes	no	yes	no	yes	no	yes	no
Varroa mites	yes	no	yes	no	yes	no	yes	no	yes	no
Other										
Overall Hive Health	1 2 3 4 5		1 2 3 4 5		1 2 3 4 5		1 2 3 4 5		1 2 3 4 5	

WHAT ACTIONS WERE TAKEN TODAY & WHAT IS TO BE SCHEDULED

HIVE INSPECTION

Date: _____ Time: _____ Temp: _____

Weather Conditions: _____

Inspector: _____ Location: _____

NAME: _____

GENERAL APPEARANCE	HIVE#		HIVE#		HIVE#		HIVE#		HIVE#	
Hive Temperment	☹	☺	☹	☺	☹	☺	☹	☺	☹	☺
Bees actively entering and exiting the hive?	yes	no	yes	no	yes	no	yes	no	yes	no
Are bees bringing pollen?	yes	no	yes	no	yes	no	yes	no	yes	no
Are there signs of disturbance?	yes	no	yes	no	yes	no	yes	no	yes	no
How many boxes?										

OPENING THE HIVES

Smoking material										
Capped honey	yes	no	yes	no	yes	no	yes	no	yes	no
Uncapped honey	yes	no	yes	no	yes	no	yes	no	yes	no
Capped brood	yes	no	yes	no	yes	no	yes	no	yes	no
Pollen	yes	no	yes	no	yes	no	yes	no	yes	no
Are there queen cells?	yes	no	yes	no	yes	no	yes	no	yes	no
Did you locate the queen?	yes	no	yes	no	yes	no	yes	no	yes	no
How many frames are covered in bees?										
Need space for nectar?	yes	no	yes	no	yes	no	yes	no	yes	no

PEST CONTROL Any signs of?

Bald brood	yes	no	yes	no	yes	no	yes	no	yes	no
Foul brood	yes	no	yes	no	yes	no	yes	no	yes	no
Bad smell	yes	no	yes	no	yes	no	yes	no	yes	no
Moths	yes	no	yes	no	yes	no	yes	no	yes	no
Mice	yes	no	yes	no	yes	no	yes	no	yes	no
Ants	yes	no	yes	no	yes	no	yes	no	yes	no
Hive beetle	yes	no	yes	no	yes	no	yes	no	yes	no
Mold	yes	no	yes	no	yes	no	yes	no	yes	no
Varroa mites	yes	no	yes	no	yes	no	yes	no	yes	no
Other										
Overall Hive Health	1 2 3 4 5		1 2 3 4 5		1 2 3 4 5		1 2 3 4 5		1 2 3 4 5	

WHAT ACTIONS WERE TAKEN TODAY & WHAT IS TO BE SCHEDULED

HIVE INSPECTION

Date: _____ Time: _____ Temp: _____

Weather Conditions: _____

Inspector: _____ Location: _____

NAME: _____

GENERAL APPEARANCE

	HIVE#		HIVE#		HIVE#		HIVE#		HIVE#	
Hive Temperment	😣	😊	😣	😊	😣	😊	😣	😊	😣	😊
Bees actively entering and exiting the hive?	yes	no	yes	no	yes	no	yes	no	yes	no
Are bees bringing pollen?	yes	no	yes	no	yes	no	yes	no	yes	no
Are there signs of disturbance?	yes	no	yes	no	yes	no	yes	no	yes	no
How many boxes?										

OPENING THE HIVES

Smoking material										
Capped honey	yes	no	yes	no	yes	no	yes	no	yes	no
Uncapped honey	yes	no	yes	no	yes	no	yes	no	yes	no
Capped brood	yes	no	yes	no	yes	no	yes	no	yes	no
Pollen	yes	no	yes	no	yes	no	yes	no	yes	no
Are there queen cells?	yes	no	yes	no	yes	no	yes	no	yes	no
Did you locate the queen?	yes	no	yes	no	yes	no	yes	no	yes	no
How many frames are covered in bees?										
Need space for nectar?	yes	no	yes	no	yes	no	yes	no	yes	no

PEST CONTROL Any signs of?

Bald brood	yes	no	yes	no	yes	no	yes	no	yes	no
Foul brood	yes	no	yes	no	yes	no	yes	no	yes	no
Bad smell	yes	no	yes	no	yes	no	yes	no	yes	no
Moths	yes	no	yes	no	yes	no	yes	no	yes	no
Mice	yes	no	yes	no	yes	no	yes	no	yes	no
Ants	yes	no	yes	no	yes	no	yes	no	yes	no
Hive beetle	yes	no	yes	no	yes	no	yes	no	yes	no
Mold	yes	no	yes	no	yes	no	yes	no	yes	no
Varroa mites	yes	no	yes	no	yes	no	yes	no	yes	no
Other										
Overall Hive Health	1 2 3 4 5		1 2 3 4 5		1 2 3 4 5		1 2 3 4 5		1 2 3 4 5	

WHAT ACTIONS WERE TAKEN TODAY & WHAT IS TO BE SCHEDULED

HIVE INSPECTION

Date: _____ Time: _____ Temp: _____

Weather Conditions: _____

Inspector: _____ Location: _____

NAME: _____

GENERAL APPEARANCE	HIVE#		HIVE#		HIVE#		HIVE#		HIVE#	
Hive Temperment	☹	☺	☹	☺	☹	☺	☹	☺	☹	☺
Bees actively entering and exiting the hive?	yes	no	yes	no	yes	no	yes	no	yes	no
Are bees bringing pollen?	yes	no	yes	no	yes	no	yes	no	yes	no
Are there signs of disturbance?	yes	no	yes	no	yes	no	yes	no	yes	no
How many boxes?										

OPENING THE HIVES										
Smoking material										
Capped honey	yes	no	yes	no	yes	no	yes	no	yes	no
Uncapped honey	yes	no	yes	no	yes	no	yes	no	yes	no
Capped brood	yes	no	yes	no	yes	no	yes	no	yes	no
Pollen	yes	no	yes	no	yes	no	yes	no	yes	no
Are there queen cells?	yes	no	yes	no	yes	no	yes	no	yes	no
Did you locate the queen?	yes	no	yes	no	yes	no	yes	no	yes	no
How many frames are covered in bees?										
Need space for nectar?	yes	no	yes	no	yes	no	yes	no	yes	no

PEST CONTROL Any signs of?										
Bald brood	yes	no	yes	no	yes	no	yes	no	yes	no
Foul brood	yes	no	yes	no	yes	no	yes	no	yes	no
Bad smell	yes	no	yes	no	yes	no	yes	no	yes	no
Moths	yes	no	yes	no	yes	no	yes	no	yes	no
Mice	yes	no	yes	no	yes	no	yes	no	yes	no
Ants	yes	no	yes	no	yes	no	yes	no	yes	no
Hive beetle	yes	no	yes	no	yes	no	yes	no	yes	no
Mold	yes	no	yes	no	yes	no	yes	no	yes	no
Varroa mites	yes	no	yes	no	yes	no	yes	no	yes	no
Other										
Overall Hive Health	1 2 3 4 5		1 2 3 4 5		1 2 3 4 5		1 2 3 4 5		1 2 3 4 5	

WHAT ACTIONS WERE TAKEN TODAY & WHAT IS TO BE SCHEDULED

HIVE INSPECTION

Date: _____ Time: _____ Temp: _____

Weather Conditions: _____

Inspector: _____

Location: _____

NAME: _____

GENERAL APPEARANCE	HIVE#		HIVE#		HIVE#		HIVE#		HIVE#	
Hive Temperment	☹	☺	☹	☺	☹	☺	☹	☺	☹	☺
Bees actively entering and exiting the hive?	yes	no	yes	no	yes	no	yes	no	yes	no
Are bees bringing pollen?	yes	no	yes	no	yes	no	yes	no	yes	no
Are there signs of disturbance?	yes	no	yes	no	yes	no	yes	no	yes	no
How many boxes?										
OPENING THE HIVES										
Smoking material										
Capped honey	yes	no	yes	no	yes	no	yes	no	yes	no
Uncapped honey	yes	no	yes	no	yes	no	yes	no	yes	no
Capped brood	yes	no	yes	no	yes	no	yes	no	yes	no
Pollen	yes	no	yes	no	yes	no	yes	no	yes	no
Are there queen cells?	yes	no	yes	no	yes	no	yes	no	yes	no
Did you locate the queen?	yes	no	yes	no	yes	no	yes	no	yes	no
How many frames are covered in bees?										
Need space for nectar?	yes	no	yes	no	yes	no	yes	no	yes	no
PEST CONTROL Any signs of?										
Bald brood	yes	no	yes	no	yes	no	yes	no	yes	no
Foul brood	yes	no	yes	no	yes	no	yes	no	yes	no
Bad smell	yes	no	yes	no	yes	no	yes	no	yes	no
Moths	yes	no	yes	no	yes	no	yes	no	yes	no
Mice	yes	no	yes	no	yes	no	yes	no	yes	no
Ants	yes	no	yes	no	yes	no	yes	no	yes	no
Hive beetle	yes	no	yes	no	yes	no	yes	no	yes	no
Mold	yes	no	yes	no	yes	no	yes	no	yes	no
Varroa mites	yes	no	yes	no	yes	no	yes	no	yes	no
Other										
Overall Hive Health	1 2 3 4 5		1 2 3 4 5		1 2 3 4 5		1 2 3 4 5		1 2 3 4 5	

WHAT ACTIONS WERE TAKEN TODAY & WHAT IS TO BE SCHEDULED

HIVE INSPECTION

Date: Time: Temp:

Weather Conditions:

Inspector: Location:

NAME:

GENERAL APPEARANCE	HIVE#		HIVE#		HIVE#		HIVE#		HIVE#	
Hive Temperment	😠	😊	😠	😊	😠	😊	😠	😊	😠	😊
Bees actively entering and exiting the hive?	yes	no	yes	no	yes	no	yes	no	yes	no
Are bees bringing pollen?	yes	no	yes	no	yes	no	yes	no	yes	no
Are there signs of disturbance?	yes	no	yes	no	yes	no	yes	no	yes	no
How many boxes?										

OPENING THE HIVES

Smoking material										
Capped honey	yes	no	yes	no	yes	no	yes	no	yes	no
Uncapped honey	yes	no	yes	no	yes	no	yes	no	yes	no
Capped brood	yes	no	yes	no	yes	no	yes	no	yes	no
Pollen	yes	no	yes	no	yes	no	yes	no	yes	no
Are there queen cells?	yes	no	yes	no	yes	no	yes	no	yes	no
Did you locate the queen?	yes	no	yes	no	yes	no	yes	no	yes	no
How many frames are covered in bees?										
Need space for nectar?	yes	no	yes	no	yes	no	yes	no	yes	no

PEST CONTROL Any signs of?

Bald brood	yes	no	yes	no	yes	no	yes	no	yes	no
Foul brood	yes	no	yes	no	yes	no	yes	no	yes	no
Bad smell	yes	no	yes	no	yes	no	yes	no	yes	no
Moths	yes	no	yes	no	yes	no	yes	no	yes	no
Mice	yes	no	yes	no	yes	no	yes	no	yes	no
Ants	yes	no	yes	no	yes	no	yes	no	yes	no
Hive beetle	yes	no	yes	no	yes	no	yes	no	yes	no
Mold	yes	no	yes	no	yes	no	yes	no	yes	no
Varroa mites	yes	no	yes	no	yes	no	yes	no	yes	no
Other										
Overall Hive Health	1 2 3 4 5		1 2 3 4 5		1 2 3 4 5		1 2 3 4 5		1 2 3 4 5	

WHAT ACTIONS WERE TAKEN TODAY & WHAT IS TO BE SCHEDULED

HIVE INSPECTION

Date: _____ Time: _____ Temp: _____

Weather Conditions: _____

Inspector: _____

Location: _____

NAME: _____

GENERAL APPEARANCE	HIVE#		HIVE#		HIVE#		HIVE#		HIVE#	
Hive Temperment	☹	☺	☹	☺	☹	☺	☹	☺	☹	☺
Bees actively entering and exiting the hive?	yes	no	yes	no	yes	no	yes	no	yes	no
Are bees bringing pollen?	yes	no	yes	no	yes	no	yes	no	yes	no
Are there signs of disturbance?	yes	no	yes	no	yes	no	yes	no	yes	no
How many boxes?										

OPENING THE HIVES										
Smoking material										
Capped honey	yes	no	yes	no	yes	no	yes	no	yes	no
Uncapped honey	yes	no	yes	no	yes	no	yes	no	yes	no
Capped brood	yes	no	yes	no	yes	no	yes	no	yes	no
Pollen	yes	no	yes	no	yes	no	yes	no	yes	no
Are there queen cells?	yes	no	yes	no	yes	no	yes	no	yes	no
Did you locate the queen?	yes	no	yes	no	yes	no	yes	no	yes	no
How many frames are covered in bees?										
Need space for nectar?	yes	no	yes	no	yes	no	yes	no	yes	no

PEST CONTROL Any signs of?										
Bald brood	yes	no	yes	no	yes	no	yes	no	yes	no
Foul brood	yes	no	yes	no	yes	no	yes	no	yes	no
Bad smell	yes	no	yes	no	yes	no	yes	no	yes	no
Moths	yes	no	yes	no	yes	no	yes	no	yes	no
Mice	yes	no	yes	no	yes	no	yes	no	yes	no
Ants	yes	no	yes	no	yes	no	yes	no	yes	no
Hive beetle	yes	no	yes	no	yes	no	yes	no	yes	no
Mold	yes	no	yes	no	yes	no	yes	no	yes	no
Varroa mites	yes	no	yes	no	yes	no	yes	no	yes	no
Other										
Overall Hive Health	1 2 3 4 5		1 2 3 4 5		1 2 3 4 5		1 2 3 4 5		1 2 3 4 5	

WHAT ACTIONS WERE TAKEN TODAY & WHAT IS TO BE SCHEDULED

HIVE INSPECTION

Date: _____ Time: _____ Temp: _____

Weather Conditions: _____

Inspector: _____ Location: _____

NAME:

GENERAL APPEARANCE	HIVE#		HIVE#		HIVE#		HIVE#		HIVE#	
Hive Temperment	☹	☺	☹	☺	☹	☺	☹	☺	☹	☺
Bees actively entering and exiting the hive?	yes	no	yes	no	yes	no	yes	no	yes	no
Are bees bringing pollen?	yes	no	yes	no	yes	no	yes	no	yes	no
Are there signs of disturbance?	yes	no	yes	no	yes	no	yes	no	yes	no
How many boxes?										

OPENING THE HIVES

Smoking material										
Capped honey	yes	no	yes	no	yes	no	yes	no	yes	no
Uncapped honey	yes	no	yes	no	yes	no	yes	no	yes	no
Capped brood	yes	no	yes	no	yes	no	yes	no	yes	no
Pollen	yes	no	yes	no	yes	no	yes	no	yes	no
Are there queen cells?	yes	no	yes	no	yes	no	yes	no	yes	no
Did you locate the queen?	yes	no	yes	no	yes	no	yes	no	yes	no
How many frames are covered in bees?										
Need space for nectar?	yes	no	yes	no	yes	no	yes	no	yes	no

PEST CONTROL Any signs of?

Bald brood	yes	no	yes	no	yes	no	yes	no	yes	no
Foul brood	yes	no	yes	no	yes	no	yes	no	yes	no
Bad smell	yes	no	yes	no	yes	no	yes	no	yes	no
Moths	yes	no	yes	no	yes	no	yes	no	yes	no
Mice	yes	no	yes	no	yes	no	yes	no	yes	no
Ants	yes	no	yes	no	yes	no	yes	no	yes	no
Hive beetle	yes	no	yes	no	yes	no	yes	no	yes	no
Mold	yes	no	yes	no	yes	no	yes	no	yes	no
Varroa mites	yes	no	yes	no	yes	no	yes	no	yes	no
Other										
Overall Hive Health	1 2 3 4 5		1 2 3 4 5		1 2 3 4 5		1 2 3 4 5		1 2 3 4 5	

WHAT ACTIONS WERE TAKEN TODAY & WHAT IS TO BE SCHEDULED

HIVE INSPECTION

Date: _____   Time: _____   Temp: _____

Weather Conditions: _____

Inspector: _____

Location: _____

NAME: _____

GENERAL APPEARANCE	HIVE#		HIVE#		HIVE#		HIVE#		HIVE#	
Hive Temperment	☹	☺	☹	☺	☹	☺	☹	☺	☹	☺
Bees actively entering and exiting the hive?	yes	no	yes	no	yes	no	yes	no	yes	no
Are bees bringing pollen?	yes	no	yes	no	yes	no	yes	no	yes	no
Are there signs of disturbance?	yes	no	yes	no	yes	no	yes	no	yes	no
How many boxes?										

OPENING THE HIVES

Smoking material										
Capped honey	yes	no	yes	no	yes	no	yes	no	yes	no
Uncapped honey	yes	no	yes	no	yes	no	yes	no	yes	no
Capped brood	yes	no	yes	no	yes	no	yes	no	yes	no
Pollen	yes	no	yes	no	yes	no	yes	no	yes	no
Are there queen cells?	yes	no	yes	no	yes	no	yes	no	yes	no
Did you locate the queen?	yes	no	yes	no	yes	no	yes	no	yes	no
How many frames are covered in bees?										
Need space for nectar?	yes	no	yes	no	yes	no	yes	no	yes	no

PEST CONTROL Any signs of?

Bald brood	yes	no	yes	no	yes	no	yes	no	yes	no
Foul brood	yes	no	yes	no	yes	no	yes	no	yes	no
Bad smell	yes	no	yes	no	yes	no	yes	no	yes	no
Moths	yes	no	yes	no	yes	no	yes	no	yes	no
Mice	yes	no	yes	no	yes	no	yes	no	yes	no
Ants	yes	no	yes	no	yes	no	yes	no	yes	no
Hive beetle	yes	no	yes	no	yes	no	yes	no	yes	no
Mold	yes	no	yes	no	yes	no	yes	no	yes	no
Varroa mites	yes	no	yes	no	yes	no	yes	no	yes	no
Other										
Overall Hive Health	1 2 3 4 5		1 2 3 4 5		1 2 3 4 5		1 2 3 4 5		1 2 3 4 5	

WHAT ACTIONS WERE TAKEN TODAY & WHAT IS TO BE SCHEDULED

HIVE INSPECTION

Date: _____ Time: _____ Temp: _____

Weather Conditions: _____

Inspector: _____

Location: _____

NAME: _____

GENERAL APPEARANCE	HIVE#		HIVE#		HIVE#		HIVE#		HIVE#	
Hive Temperment	☹	☺	☹	☺	☹	☺	☹	☺	☹	☺
Bees actively entering and exiting the hive?	yes	no	yes	no	yes	no	yes	no	yes	no
Are bees bringing pollen?	yes	no	yes	no	yes	no	yes	no	yes	no
Are there signs of disturbance?	yes	no	yes	no	yes	no	yes	no	yes	no
How many boxes?										
OPENING THE HIVES										
Smoking material										
Capped honey	yes	no	yes	no	yes	no	yes	no	yes	no
Uncapped honey	yes	no	yes	no	yes	no	yes	no	yes	no
Capped brood	yes	no	yes	no	yes	no	yes	no	yes	no
Pollen	yes	no	yes	no	yes	no	yes	no	yes	no
Are there queen cells?	yes	no	yes	no	yes	no	yes	no	yes	no
Did you locate the queen?	yes	no	yes	no	yes	no	yes	no	yes	no
How many frames are covered in bees?										
Need space for nectar?	yes	no	yes	no	yes	no	yes	no	yes	no
PEST CONTROL Any signs of?										
Bald brood	yes	no	yes	no	yes	no	yes	no	yes	no
Foul brood	yes	no	yes	no	yes	no	yes	no	yes	no
Bad smell	yes	no	yes	no	yes	no	yes	no	yes	no
Moths	yes	no	yes	no	yes	no	yes	no	yes	no
Mice	yes	no	yes	no	yes	no	yes	no	yes	no
Ants	yes	no	yes	no	yes	no	yes	no	yes	no
Hive beetle	yes	no	yes	no	yes	no	yes	no	yes	no
Mold	yes	no	yes	no	yes	no	yes	no	yes	no
Varroa mites	yes	no	yes	no	yes	no	yes	no	yes	no
Other										
Overall Hive Health	1 2 3 4 5		1 2 3 4 5		1 2 3 4 5		1 2 3 4 5		1 2 3 4 5	

WHAT ACTIONS WERE TAKEN TODAY & WHAT IS TO BE SCHEDULED

_____

_____

_____

_____

_____

_____

_____

_____

_____

HIVE INSPECTION

Date: _____ Time: _____ Temp: _____

Weather Conditions: _____

Inspector: _____ Location: _____

NAME: _____

GENERAL APPEARANCE	HIVE#		HIVE#		HIVE#		HIVE#		HIVE#	
Hive Temperment	☹	☺	☹	☺	☹	☺	☹	☺	☹	☺
Bees actively entering and exiting the hive?	yes	no	yes	no	yes	no	yes	no	yes	no
Are bees bringing pollen?	yes	no	yes	no	yes	no	yes	no	yes	no
Are there signs of disturbance?	yes	no	yes	no	yes	no	yes	no	yes	no
How many boxes?										

OPENING THE HIVES										
Smoking material										
Capped honey	yes	no	yes	no	yes	no	yes	no	yes	no
Uncapped honey	yes	no	yes	no	yes	no	yes	no	yes	no
Capped brood	yes	no	yes	no	yes	no	yes	no	yes	no
Pollen	yes	no	yes	no	yes	no	yes	no	yes	no
Are there queen cells?	yes	no	yes	no	yes	no	yes	no	yes	no
Did you locate the queen?	yes	no	yes	no	yes	no	yes	no	yes	no
How many frames are covered in bees?										
Need space for nectar?	yes	no	yes	no	yes	no	yes	no	yes	no

PEST CONTROL Any signs of?										
Bald brood	yes	no	yes	no	yes	no	yes	no	yes	no
Foul brood	yes	no	yes	no	yes	no	yes	no	yes	no
Bad smell	yes	no	yes	no	yes	no	yes	no	yes	no
Moths	yes	no	yes	no	yes	no	yes	no	yes	no
Mice	yes	no	yes	no	yes	no	yes	no	yes	no
Ants	yes	no	yes	no	yes	no	yes	no	yes	no
Hive beetle	yes	no	yes	no	yes	no	yes	no	yes	no
Mold	yes	no	yes	no	yes	no	yes	no	yes	no
Varroa mites	yes	no	yes	no	yes	no	yes	no	yes	no
Other										
Overall Hive Health	1 2 3 4 5		1 2 3 4 5		1 2 3 4 5		1 2 3 4 5		1 2 3 4 5	

WHAT ACTIONS WERE TAKEN TODAY & WHAT IS TO BE SCHEDULED

_____

_____

_____

_____

_____

_____

_____

HIVE INSPECTION

Date: _____ Time: _____ Temp: _____

Weather Conditions: _____

Inspector: _____ Location: _____

NAME:

GENERAL APPEARANCE	HIVE#		HIVE#		HIVE#		HIVE#		HIVE#	
Hive Temperment	☹	☺	☹	☺	☹	☺	☹	☺	☹	☺
Bees actively entering and exiting the hive?	yes	no	yes	no	yes	no	yes	no	yes	no
Are bees bringing pollen?	yes	no	yes	no	yes	no	yes	no	yes	no
Are there signs of disturbance?	yes	no	yes	no	yes	no	yes	no	yes	no
How many boxes?										

OPENING THE HIVES										
Smoking material										
Capped honey	yes	no	yes	no	yes	no	yes	no	yes	no
Uncapped honey	yes	no	yes	no	yes	no	yes	no	yes	no
Capped brood	yes	no	yes	no	yes	no	yes	no	yes	no
Pollen	yes	no	yes	no	yes	no	yes	no	yes	no
Are there queen cells?	yes	no	yes	no	yes	no	yes	no	yes	no
Did you locate the queen?	yes	no	yes	no	yes	no	yes	no	yes	no
How many frames are covered in bees?										
Need space for nectar?	yes	no	yes	no	yes	no	yes	no	yes	no

PEST CONTROL Any signs of?										
Bald brood	yes	no	yes	no	yes	no	yes	no	yes	no
Foul brood	yes	no	yes	no	yes	no	yes	no	yes	no
Bad smell	yes	no	yes	no	yes	no	yes	no	yes	no
Moths	yes	no	yes	no	yes	no	yes	no	yes	no
Mice	yes	no	yes	no	yes	no	yes	no	yes	no
Ants	yes	no	yes	no	yes	no	yes	no	yes	no
Hive beetle	yes	no	yes	no	yes	no	yes	no	yes	no
Mold	yes	no	yes	no	yes	no	yes	no	yes	no
Varroa mites	yes	no	yes	no	yes	no	yes	no	yes	no
Other										
Overall Hive Health	1 2 3 4 5		1 2 3 4 5		1 2 3 4 5		1 2 3 4 5		1 2 3 4 5	

WHAT ACTIONS WERE TAKEN TODAY & WHAT IS TO BE SCHEDULED

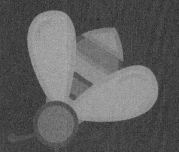

MONTHLY SCHEDULE

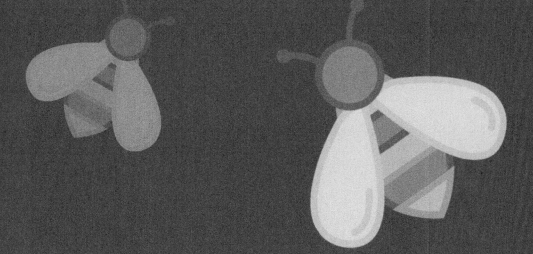

MONTHLY SCHEDULE

MONTH:

SUNDAY	MONDAY	TUESDAY	WEDNESDAY	THURSDAY	FRIDAY	SATURDAY
☐	☐	☐	☐	☐	☐	☐
☐	☐	☐	☐	☐	☐	☐
☐	☐	☐	☐	☐	☐	☐
☐	☐	☐	☐	☐	☐	☐
☐	☐	☐	☐	☐	☐	☐

MONTHLY SCHEDULE

MONTH:

SUNDAY	MONDAY	TUESDAY	WEDNESDAY	THURSDAY	FRIDAY	SATURDAY

MONTHLY SCHEDULE

MONTH:

SUNDAY	MONDAY	TUESDAY	WEDNESDAY	THURSDAY	FRIDAY	SATURDAY

MONTHLY SCHEDULE

MONTH: _____

SUNDAY	MONDAY	TUESDAY	WEDNESDAY	THURSDAY	FRIDAY	SATURDAY

MONTHLY SCHEDULE

MONTH:

SUNDAY	MONDAY	TUESDAY	WEDNESDAY	THURSDAY	FRIDAY	SATURDAY

MONTHLY SCHEDULE

MONTH: _____

SUNDAY	MONDAY	TUESDAY	WEDNESDAY	THURSDAY	FRIDAY	SATURDAY
☐	☐	☐	☐	☐	☐	☐
☐	☐	☐	☐	☐	☐	☐
☐	☐	☐	☐	☐	☐	☐
☐	☐	☐	☐	☐	☐	☐
☐	☐	☐	☐	☐	☐	☐

MONTHLY SCHEDULE

MONTH: _____

SUNDAY	MONDAY	TUESDAY	WEDNESDAY	THURSDAY	FRIDAY	SATURDAY

MONTHLY SCHEDULE

MONTH:

SUNDAY	MONDAY	TUESDAY	WEDNESDAY	THURSDAY	FRIDAY	SATURDAY

MONTHLY SCHEDULE

MONTH: _____

SUNDAY	MONDAY	TUESDAY	WEDNESDAY	THURSDAY	FRIDAY	SATURDAY

MONTHLY SCHEDULE

MONTH: _____

SUNDAY	MONDAY	TUESDAY	WEDNESDAY	THURSDAY	FRIDAY	SATURDAY

SUNDAY	MONDAY	TUESDAY	WEDNESDAY	THURSDAY	FRIDAY	SATURDAY

MONTHLY SCHEDULE

MONTH:

SUNDAY	MONDAY	TUESDAY	WEDNESDAY	THURSDAY	FRIDAY	SATURDAY
☐	☐	☐	☐	☐	☐	☐
☐	☐	☐	☐	☐	☐	☐
☐	☐	☐	☐	☐	☐	☐
☐	☐	☐	☐	☐	☐	☐
☐	☐	☐	☐	☐	☐	☐

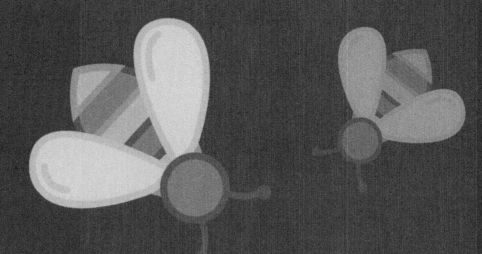

Contacts
References

REFERENCES

Here's a place to list great ideas, organizations, youtube channels, podcasts, websites, blogs and anything that's worh sharing.

REFERENCES

Here's a place to list great ideas, organizations, youtube channels, podcasts, websites, blogs and anything that's worh sharing.

CONTACTS

List your important contacts, like your customers, suppliers, organizations and fellow bee keepers.

NAME: _____   BUSINESS: _____

TEL: _____   EMAIL: _____

ADDRESS: _____   WEBSITE: _____

NAME: _____   BUSINESS: _____

TEL: _____   EMAIL: _____

ADDRESS: _____   WEBSITE: _____

NAME: _____   BUSINESS: _____

TEL: _____   EMAIL: _____

ADDRESS: _____   WEBSITE: _____

NAME: _____   BUSINESS: _____

TEL: _____   EMAIL: _____

ADDRESS: _____   WEBSITE: _____

NAME: _____   BUSINESS: _____

TEL: _____   EMAIL: _____

ADDRESS: _____   WEBSITE: _____

NAME: _____   BUSINESS: _____

TEL: _____   EMAIL: _____

ADDRESS: _____   WEBSITE: _____

NAME: _____   BUSINESS: _____

TEL: _____   EMAIL: _____

ADDRESS: _____   WEBSITE: _____

NAME: _____   BUSINESS: _____

TEL: _____   EMAIL: _____

ADDRESS: _____   WEBSITE: _____

CONTACTS

List your important contacts, like your customers, suppliers, organizations and fellow bee keepers.

NAME: ....

TEL: ....

ADDRESS: ....

BUSINESS: ....

EMAIL: ....

WEBSITE: ....

NAME: ....

TEL: ....

ADDRESS: ....

BUSINESS: ....

EMAIL: ....

WEBSITE: ....

NAME: ....

TEL: ....

ADDRESS: ....

BUSINESS: ....

EMAIL: ....

WEBSITE: ....

NAME: ....

TEL: ....

ADDRESS: ....

BUSINESS: ....

EMAIL: ....

WEBSITE: ....

NAME: ....

TEL: ....

ADDRESS: ....

BUSINESS: ....

EMAIL: ....

WEBSITE: ....

NAME: ....

TEL: ....

ADDRESS: ....

BUSINESS: ....

EMAIL: ....

WEBSITE: ....

NAME: ....

TEL: ....

ADDRESS: ....

BUSINESS: ....

EMAIL: ....

WEBSITE: ....

NAME: ....

TEL: ....

ADDRESS: ....

BUSINESS: ....

EMAIL: ....

WEBSITE: ....

CONTACTS

List your important contacts, like your customers, suppliers, organizations and fellow bee keepers.

NAME: .... BUSINESS: ....
TEL: .... EMAIL: ....
ADDRESS: .... WEBSITE: ....

NAME: .... BUSINESS: ....
TEL: .... EMAIL: ....
ADDRESS: .... WEBSITE: ....

NAME: .... BUSINESS: ....
TEL: .... EMAIL: ....
ADDRESS: .... WEBSITE: ....

NAME: .... BUSINESS: ....
TEL: .... EMAIL: ....
ADDRESS: .... WEBSITE: ....

NAME: .... BUSINESS: ....
TEL: .... EMAIL: ....
ADDRESS: .... WEBSITE: ....

NAME: .... BUSINESS: ....
TEL: .... EMAIL: ....
ADDRESS: .... WEBSITE: ....

NAME: .... BUSINESS: ....
TEL: .... EMAIL: ....
ADDRESS: .... WEBSITE: ....

NAME: .... BUSINESS: ....
EL: .... EMAIL: ....
ADDRESS: .... WEBSITE: ....

CONTACTS

List your important contacts, like your customers, suppliers, organizations and fellow bee keepers.

NAME: BUSINESS:

TEL: EMAIL:

ADDRESS: WEBSITE:

NAME: BUSINESS:

TEL: EMAIL:

ADDRESS: WEBSITE:

NAME: BUSINESS:

TEL: EMAIL:

ADDRESS: WEBSITE:

NAME: BUSINESS:

TEL: EMAIL:

ADDRESS: WEBSITE:

NAME: BUSINESS:

TEL: EMAIL:

ADDRESS: WEBSITE:

NAME: BUSINESS:

TEL: EMAIL:

ADDRESS: WEBSITE:

NAME: BUSINESS:

TEL: EMAIL:

ADDRESS: WEBSITE:

NOTES

BEE NOTES

BEE NOTES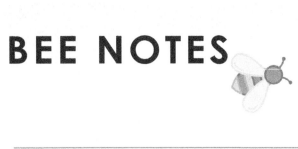

BEE NOTES

BEE NOTES

BEE NOTES

BEE NOTES

BEE NOTES

BEE NOTES

BEE NOTES

BEE NOTES

BEE NOTES

BEE NOTES

BEE NOTES

Made in United States
Orlando, FL
20 April 2024

45980599R00063